T0292393

Springer Theses

Recognizing Outstanding Ph.D. Research

Aims and Scope

The series "Springer Theses" brings together a selection of the very best Ph.D. theses from around the world and across the physical sciences. Nominated and endorsed by two recognized specialists, each published volume has been selected for its scientific excellence and the high impact of its contents for the pertinent field of research. For greater accessibility to non-specialists, the published versions include an extended introduction, as well as a foreword by the student's supervisor explaining the special relevance of the work for the field. As a whole, the series will provide a valuable resource both for newcomers to the research fields described, and for other scientists seeking detailed background information on special questions. Finally, it provides an accredited documentation of the valuable contributions made by today's younger generation of scientists.

Theses are accepted into the series by invited nomination only and must fulfill all of the following criteria

- They must be written in good English.
- The topic should fall within the confines of Chemistry, Physics, Earth Sciences, Engineering and related interdisciplinary fields such as Materials, Nanoscience, Chemical Engineering, Complex Systems and Biophysics.
- The work reported in the thesis must represent a significant scientific advance.
- If the thesis includes previously published material, permission to reproduce this must be gained from the respective copyright holder.
- They must have been examined and passed during the 12 months prior to nomination.
- Each thesis should include a foreword by the supervisor outlining the significance of its content.
- The theses should have a clearly defined structure including an introduction accessible to scientists not expert in that particular field.

More information about this series at http://www.springer.com/series/8790

Khalid Karam Abd

Intelligent Scheduling of Robotic Flexible Assembly Cells

Doctoral Thesis accepted by
University of South Australia, Adelaide, Australia

 Springer

Author
Dr. Khalid Karam Abd
School of Engineering
University of South Australia
Adelaide, SA
Australia

Supervisors
Prof. Kazem Abhary
School of Engineering
University of South Australia
Adelaide
Australia

Dr. Romeo Marian
School of Engineering
University of South Australia
Adelaide
Australia

ISSN 2190-5053   ISSN 2190-5061   (electronic)
Springer Theses
ISBN 978-3-319-26295-6   ISBN 978-3-319-26296-3   (eBook)
DOI 10.1007/978-3-319-26296-3

Library of Congress Control Number: 2015954968

Springer Cham Heidelberg New York Dordrecht London

Printed on acid-free paper

Springer International Publishing AG Switzerland is part of Springer Science+Business Media (www.springer.com)

Parts of this thesis have been published in the following documents:

Journals

Abd, K, Abhary, K & Marian, R 2014, 'Simulation modelling and analysis of scheduling in robotic flexible assembly cells using Taguchi method', *International Journal of Production Research*, vol. 52, no. 12, pp. 2654–2666.

Abd, K, Abhary, K & Marian, R 2013, 'Fuzzy decision support system for selecting the optimal scheduling rule in robotic flexible assembly cells', *Australian Journal of Multi-Disciplinary Engineering*, vol. 9, no. 2, pp. 125–132.

Abd, K, Abhary, K & Marian, R 2013, 'Application of fuzzy logic to multi-objective scheduling problems in robotic flexible assembly cells', *Automation, Control and Intelligent Systems*, vol. 1, no. 3, pp. 34–41.

Abd, K, Abhary, K & Marian, R 2013, 'Development of a fuzzy-simulation model of scheduling robotic flexible assembly cells', *Journal of Computer Science*, vol. 9, no. 12, pp. 1761–1768.

Abd, K, Abhary, K & Marian, R 2013, 'Application of a fuzzy-simulation model of scheduling robotic flexible assembly cells', *Journal of Computer Science*, vol. 9, no. 12, pp. 1769–1777.

Abd, K, Abhary, K & Marian, R 2011, 'Scheduling and performance evaluation of robotic flexible assembly cells under different dispatching rules', *Advances in Mechanical Engineering*, vol. 1, no. 1, pp. 31–40.

Abd, K, Abhary, K & Marian, R 2011, 'An MCDM approach to selection scheduling rule in robotic flexible assembly cells', *World Academy of Science, Engineering and Technology*, vol. 76, no. 1, pp. 643–648.

Abd, K, Abhary, K & Marian, R 2010, 'A Scheduling framework for robotic flexible assembly cells', *AIJSTPME-Asian International Journal of Science and Technology in Production and Manufacturing Engineering*, vol. 4, no. 1, pp. 30–37.

International Conferences

Abd, K, Abhary, K & Marian, R, 2014, 'A methodology for fuzzy multi-criteria decision-making approach for scheduling problems in robotic flexible assembly cells', accepted for publication in *the IEEE International Conference on Industrial Engineering and Engineering Management*, Malaysia, 9–12 December 2014.

Abd, K, Abhary, K & Marian, R, 2014 'Application of a fuzzy multi-criteria decision-making approach for scheduling problems in robotic flexible assembly cells', accepted for publication in *the IEEE International Conference on Industrial Engineering and Engineering Management*, Malaysia, 9–12 December 2014.

Abd, K, Abhary, K & Marian, R 2013, 'Comparative analysis of MCDM methods and implementation of the scheduling rule selection problem: a case study in robotic flexible assembly cells', *International Conference on Advanced Engineering Optimization Through Intelligent Techniques*, India, pp. 16–20.

Abd, K, Abhary, K & Marian, R 2013, 'A methodology for scheduling robotic flexible assembly cells using fuzzy logic and simulation', *Proceedings of the World Congress on Engineering*, London, United Kingdom, pp. 449–454.

Abd, K, Abhary, K & Marian, R 2012, 'Efficient scheduling rule for robotic flexible assembly cells based on fuzzy approach', *45th CIRP Conference on Manufacturing Systems*, Athens, Greece, pp. 483–488.

Abd, K, Abhary, K & Marian, R 2012, 'Intelligent modeling of scheduling robotic flexible assembly cells using fuzzy logic', *12th WSEAS International Conference on Robotics, Control and Manufacturing Technology*, Rovaniemi, Finland, pp. 202–207.

Book Chapters

Abd, K, Abhary, K & Marian, R 2014, 'Development of an intelligent approach to dynamic scheduling in robotic flexible assembly cells', *IAENG Transaction on Engineering Science*, Taylor and Francis Group, London, pp. 203–214.

Abd, K, Abhary, K & Marian, R 2013, 'Intelligent model of scheduling RFAC—Part I: methodology and strategy', *DAAAM International Scientific Book* Vienna: DAAAM International Publishing, pp. 719–736.

Abd, K, Abhary, K & Marian, R 2013, 'Intelligent model of scheduling RFAC—Part II: application', *DAAAM International Scientific Book* Vienna: DAAAM International Publishing, pp. 737–750.

Supervisor's Foreword

Flexibility signifies a manufacturing system's ability to respond quickly and effectively to market changes which are often unpredictable and imply the need for increasingly flexible systems. One such system is the robotic flexible assembly cell (RFAC). The design of RFAC with multirobots leads to increased productivity. However, more than one robot operating simultaneously in the same work environment requires a complex scheduling system to prevent collisions between robots in the shared area.

Dr. Abd who started his research with critical review of the literature on scheduling problems in RFAC revealed that only few studies have been done on scheduling RFAC and that these research studies have been limited to three major issues: (1) single-product assembly environments; (2) static situations without considering any dynamic events; and (3) nearly all of them confined to solving single-objective optimization problems.

Dr. Abd's research fills this huge knowledge gap by presenting a novel strategy for addressing the scheduling problems in RFAC. Due to the complexity of developing this strategy, his research is divided into three steps. The first step is to develop a new methodology for scheduling RFAC in a multiproduct assembly environment. The second step is to develop an intelligent approach for scheduling RFAC in dynamic situations. The last step is to extend the developed approach to optimize the dynamic scheduling of RFAC for multiobjective problems. These three steps are performed using a combination of advanced solution approaches such as simulation, fuzzy logic, system modeling, and Taguchi optimization method.

Through a sufficiently complex case study in this thesis, Dr. Abd's shows that the developed strategy is practical and effectively achieves the desired outcome. In addition to significantly contributing toward solving scheduling problems of RFAC, his research opens areas for future research in order to complement this research and make it even more applicable to real-world problems.

Supervising Dr. Abd was quite pleasing for me. He was an exceptional Ph.D. student. He worked very hard with joy and perseverance. Among many Ph.D.

students, I have supervised that he has been the most productive one evidenced by publishing eight refereed journal papers, six refereed international conference papers, and three book chapters during his Ph.D. candidateship. From this point of view, he produced the most prominent Ph.D. thesis in my school in 2014.

Adelaide   Prof. Kazem Abhary
November 2015

Acknowledgments

First and foremost, all praises are due to Allah, the Almighty, beneficent, and merciful. I gratefully thank him, who has given me the strength to achieve this work.

I would like to express my deepest gratitude to my supervisors, Professor *Kazem Abhary* and Dr. *Romeo Marian*, for their invaluable advice, endless support, and guidance to fulfill this work. Without them, it would not have been possible for this thesis to become a reality. Working under their supervision was a great learning experience not just to accomplish a doctoral thesis but also to become a successful researcher in the future. I would like to thank them for allowing me to grow as a research scientist.

I would like to thank the School of Engineering at the University of South Australia (UniSA) which has provided me with a comfortable and quiet environment during my research. UniSA also funded me to attend various national and international conferences, and I really appreciate and thank the university for that. Many thanks are extended to the School of Engineering's staff. In particular, I would like to thank *Danielle Richardson, Elizabeth Csavas, Maria Chuang*, and *Cho Lee* for their valuable support and assistance during my research study.

No words of gratitude are enough to thank Dr. *Monica Behrend* and Dr. *Judy Ford* for their effort in running the countless workshops for Ph.D. students. These workshops have really developed my written and oral communication skills. Thanks also to the library staff, especially Learning and Teaching Unit staff for their sincere advices.

I would like to express my sincere appreciation and thanks to my proofreader Ms. *Julie Piesiewicz*. I hope she knows how much I appreciate the help and support she gave me during more than a year. Also, a special thank you to my neighbor, Mr. *Heinz Schulz*, for his ongoing support to me and my family during our stay in Adelaide.

I would also like to thank the Iraqi Government, especially the Ministry of Higher Education and Scientific Research for the scholarship support to pursue my doctoral studies at the University of South Australia.

Last but not least, I would like to thank my family, colleagues, and friends here in Adelaide and in my home country for their encouragement and support during my Ph.D. journey.

Contents

Abbreviations

ACO	Ant colony optimization
AHP	Analytic hierarchy process
AI	Artificial intelligence
ANOM	Analysis of mean
ANOVA	Analysis of variance
CNC	Computer numerical control
CON	Constant
DR	Dispatching rules
FAHP	Fuzzy analytic hierarchy process
FAS	Flexible assembly systems
FDSS	Fuzzy decision support system
FL	Fuzzy logic
FLS	Fuzzy logic system
FMS	Flexible manufacturing (machining) systems
FPR	Fuzzy priority rule
FSR	Fuzzy sequencing rule
FTOPSIS	Fuzzy technique for order preference by similarity to ideal solution
GA	Genetic algorithms
GT	Group technology
K	Due-date tightness factor
MADM	Multiattribute decision making
MCDA	Multicriteria decision analysis
MCDM	Multicriteria decision making
MF	Membership function
MODS	Multiobjective decision support
MPCI	Multiple performance characteristics index
NN	Neural networks
NOP	Number of operations
NP-hard	Nondeterministic polynomial-time hard
PCB	Printed circuit board
RAC	Robotic assembly cell

RAL	Robotic assembly line
RAN	Random
RFAC	Robotic flexible assembly cell
RPD	Relative percentage deviation
S/N	Signal to noise
SLK	Slack
SMART	Simple multiattribute rating technique
SR	Sequencing rules
TFN	Triangular fuzzy number
TOPSIS	Technique for order preference by similarity to ideal solution
TrFN	Trapezoidal fuzzy numbers
TS	Tabu search
TWK	Total work content
U	Cell utilization
WPM	Weighted product model
WSM	Weighted sum model

Chapter 1
Background and Research Scope

Abstract With the rapidly developing global economy, today's companies face greater challenges than ever to employ manufacturing systems capable of dealing with unexpected events and meeting customers' specific requirements. In overcoming these challenges, flexibility is the key concept in the development of manufacturing systems. The issue of flexibility in manufacturing systems is not new and has attracted significant attention by researchers since the development of flexible manufacturing systems (FMS) four decades ago.

1.1 Introduction

With the rapidly developing global economy, today's companies face greater challenges than ever to employ manufacturing systems capable of dealing with unexpected events and meeting customers' specific requirements. In overcoming these challenges, flexibility is the key concept in the development of manufacturing systems. The issue of flexibility in manufacturing systems is not new and has attracted significant attention by researchers since the development of flexible manufacturing systems (FMS) four decades ago. The introduction of FMS in industry has provided significant potential benefit both for reducing production time and for responding to unpredictable market demands (Udhayakumar and Kumanan 2010; Joseph and Sridharan 2011).

FMS are typically classified into two main subsets: flexible assembly systems (FAS) and flexible machining systems (FMS) (Browne et al. 1984; Maccarthy and Liu 1993). Much of the research work has focused on FMS, whereas FAS have received less attention from researchers. Both FAS and FMS can be generally described as computer integrated manufacturing systems. Nevertheless, they differ in several respects (Lee et al. 2006): first, the number of different tasks performed in FAS is much more than that in FMS. In FAS, several components are jointed at the same time, while FMS involve operating only one part at a time; second, the

© Springer International Publishing Switzerland 2016
K.K. Abd, *Intelligent Scheduling of Robotic Flexible Assembly Cells*,
Springer Theses, DOI 10.1007/978-3-319-26296-3_1

processing time of each operation in FAS is much less than the time required in FMS. Therefore, the ratio of the transfer time to the processing time in FAS is high compared with that in FMS; third, the material handling system in FAS is more complicated in comparison with that in FMS. These differences make the decision problems in FAS more difficult compared with those in FMS. The purpose of this introductory chapter is:

- To describe flexible assembly systems (FAS) and the two main types of FAS, robotic assembly line (RAL) and robotic assembly cell (RAC), and to highlight the advantageous features of the RAC.
- To briefly outline the important issues in the design, planning, scheduling and controlling of RAC, and then clarify the significance of studying the scheduling problems in RAC.
- To summarize the motivations that led to this research.
- To identify the scope of the research and the main aim of this thesis.

1.2   Flexible Assembly Systems

A flexible assembly system (FAS) is a completely integrated system which consists of a number of stations that can assemble different part types and are connected by an automated material handling system and controlled by a central computer (Sawik 1998; Zhanga et al. 2005). The main components of FAS can be classified into two types: robots and peripheral equipment. These components are presented in the following sub-sections.

1.2.1   Robots

Robots are considered as crucial components in the assembly systems (Levitin et al. 2006). In general, robots fetch the assembled parts and move them from one station to another. Different types of assembly robots have different efficiencies and capabilities for the various elements of assembly tasks (Gao et al. 2009). Assembly robots can be categorized into two main types: Selective Compliant Assembly Robot Arm (SCARA) and Articulated Robots (Groover 2008).

SCARA is considered as one of the most popular robots for automatic assembly. A SCARA is rigid in the Z-axis and has full range of motion on its X and Y axes. This robot design gives four axes freedom of movement, as shown in Fig. 1.1. SCARA has high speed for vertical assembly and is suitable for assembling small parts such as electromechanical components (Taylan and Canan 2005; Salman et al. 2009) in one set direction (along Z axis).

Articulated robots or industrial robots have six axes and consequently more freedom of movement, as shown in Fig. 1.2. For this reason, articulated robots are

Fig. 1.1 Four-axis SCARA
robots

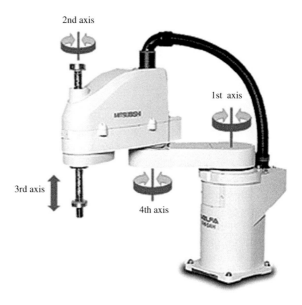

Fig. 1.2 Six-axis articulated
robot

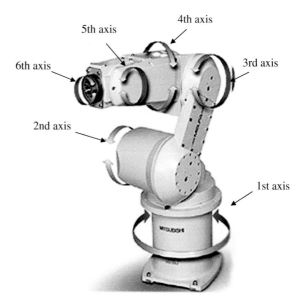

the key solution for improved flexibility and increased productivity in industrial
systems (Pan et al. 2012). Due to the increasing trend of using this robot, recent
research has been devoted to making articulated robots easy to use (Qi et al. 2008;
Zhang and Qi 2008).

1.2.2 Peripheral Equipment

FAS generally consist of different peripheral equipment for assembly operations. Peripheral equipment can be divided into five essential types (Sawik 1999; Delchambre 1992):

(1) Assembly stations where the parts are assembled.
(2) A gripper changing station where the grippers are changed.
(3) Material handling devices such as input conveyors for supplying the base parts, and output conveyors for conveying out a final product when assembly processes are completed.
(4) Storage areas such as tables for subassemblies and part feeders for supplying parts to the cells.
(5) Accessories for assembly operations such as tools for executing a fastening process, grippers for transferring and positioning the parts or subassemblies, and fixtures for holding the components in an assembly station during construction of a product.

1.3 Classification of Flexible Assembly Systems

The different FAS can be categorized according to many characteristics such as material flow configuration, system layout, capacity and capability of assembly machines related to large volume/limited variety and limited volume/large variety production (Sawik 2004; Rosati et al. 2013). FAS can be divided into two main types (Sawik 1999): robotic assembly line (RAL) and robotic assembly cells (RAC). These types and their characteristics are explained in the following sub-sections.

1.3.1 Robotic Assembly Line

A robotic assembly line (RAL) is a flow type system consisting of a series of special purpose robotic assembly stations connected by an automated material handling system, as shown in Fig. 1.3. A RAL is used for high volume/low variety assembly of a few products that have stable designs and demand requirements (Levitin et al. 2006; Daoud et al. 2014; Sawik 1999). A RAL can be compared with a conventional transfer line; it uses special purpose machines, and hence has the ability to achieve high productivity (Yoosefelahi et al. 2012; Gao et al. 2009; Bukchin and Tzur 2000).

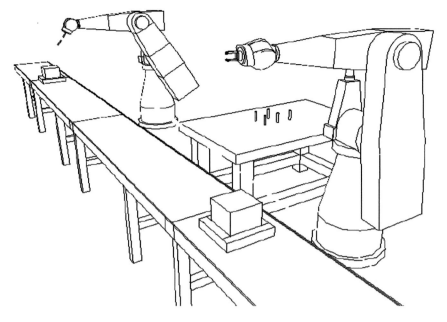

Fig. 1.3 Robotic assembly line simulation model (Cheng 2000)

1.3.2   Robotic Assembly Cell

Robotic assembly cell (RAC) is a highly modern system, structured with industrial robot(s), assembly stations and an automated material handling system, all monitored by computer numerical control (Manivannan 1993; Marian et al. 2003). RAC are capable of assembling a large variety of products in small to medium batch sizes (Mohamed et al. 2001).

The design of RAC with multi-robots has three key advantages for industrial companies. First, the RAC layout with multi-robots is very flexible and combines the productivity of product-flow layout with the flexibility of process-based layout. In the RAC, the multi-robots are used for the rapid transfer of parts and partial assemblies between highly productive assembly stations. The second advantage comes from the need of the assembly process for robots with different characteristics such as end effectors, payload, repeatability, degrees of freedom and accuracy. The third advantage is the ability of robots to employ end effectors as fixtures to allow reduction of complex orienting, because the assembly processes of products could require more than one direction of part insertion. With these advantages, employing multi-robots in the assembly cells obviously allows for increased flexibility and reconfiguration capacity (Gilbert et al. 1990) (Fig. 1.4).

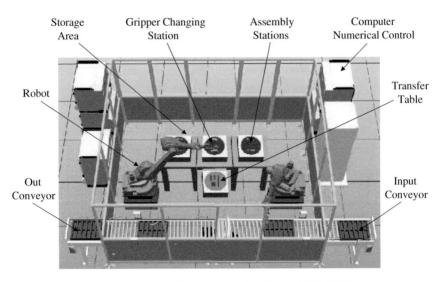

Storage   Gripper Changing   Assembly   Computer
Area   Station   Stations   Numerical Control

Robot   Transfer
                                                                        Table

Out   Input
Conveyor   Conveyor

Fig. 1.4 Robotic flexible assembly cell made by ABB software (ABB 2013)

1.3.3   Simple Comparison Between RAL and RAC

Based on the characteristics of the RAL and RAC, as shown in Table 1.1, RAC can
be considered to have higher dexterity and flexibility than RAL due to the following
reasons: (1) In RAC, the sequence of the assembly process is unconstrained, while
in RAL, only one way of assembling the product is possible; (2) RAC are easier to
modify and reconfigure and also may need less space compared with RAL;
(3) RAC are more adaptable to assembling a variety of products using the same

Table 1.1 The difference between RAL and RAC (Makino 1989)

Characteristics	RAL	RAC
Process	Divided	Integrated
Task of each station	Simplified Specialized Standardized	Complex Versatile Not standardized
Cycle time	Short (3–60 s)	Long (3–20 min)
Flow of work piece	In-line One-way	Circulating Network
Number of robots	1–100	1–4
No. of assembly parts	1–6	1–50
Tool change	Single tool Multi finger	Switchable tool Automatic tool change
Production volume	High	Low

resources; (4) In RAC, the equipment and machines are multi-purpose, whilst in RAL equipment is dedicated to the specific products. For the above reasons, it can be concluded that RAC have a higher degree of flexibility than RAL. Hence, this research will focus on the RAC. The acronym (RFAC) for *Robotic Flexible Assembly Cell* will be used in the rest of this research, as used in other relevant research.

1.4 Scheduling of Robotic Flexible Assembly Cell

Flexible assembly systems are typically fully integrated production systems. They consist of more than one robot and much peripheral equipment. For this reason, the use of such a system is extremely complex. In order to employ FAS as effectively as possible due to their complexity and the high cost of the robots, the four main stages of decision making (*design*, *planning*, *scheduling* and *control*) in the FAS should work effectively (Kazerooni 1997; Udhayakumar and Kumanan 2012). Taking into account the time horizon in flexible assembly systems, the four stages can be redescribed as three interconnected decision levels (Stecke 1985; Schneeweiss 1995), as shown in Fig. 1.5 and explained below:

- FAS *design* problems involve layout design, assembly machines and material handling selection, and product design for automated assembly. They also involve assembly planning (Bukchin and Tzur 2000; Park et al. 2001; Nakase et al. 2002). These problems generally have long-term implications (Schneeweiss 1995).
- FAS *planning* problems are medium-term problems including resource allocation, and machine loading (Schneeweiss 1995; Sawik 1999). The *planning*

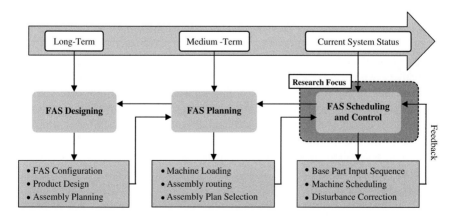

Fig. 1.5 Decision levels in FASs

problems also include assembly planning which is associated with the sequence planning to be done in order to assemble the final products (Rosell 2004).

- FAS *scheduling* and *control* are major short-term problems. The objective of *scheduling* problems is to address the detailed sequencing of all the assembly tasks that are required to assemble a product (Sawik 2004). The *control* problems are considered as part of the production planning and scheduling (Kazerooni 1997; Chan et al. 2002). The objective of control problems is to monitor the system performance and decide whether or not the system status needs corrective actions (Schneeweiss 1995; Sawik 1999).

Clossen and Malstrom (1982) stated that "hundreds of robots and millions of dollars' worth of computer-controlled equipment are worthless if they are under-utilized or if they spend their time working on the wrong part because of poor planning and scheduling". Additionally, many researchers confirm that scheduling problems play a crucial role in determining the system's performance compared with design and planning problems (Joseph and Sridharan 2011; Udhayakumar and Kumanan 2012; Burnwal and Deb 2013). Consequently, this research has mainly focused on the *scheduling* problems of robotic flexible assembly cells (RFAC), which are considered one type of FAS.

1.5 Motivation for Research in Scheduling of RFAC

The design of RFAC with multi-robots leads to increased productivity in a shorter cycle time and with lower production costs. However, there are certain difficulties that have arisen with this design concept. For example, a system with more than one robot operating simultaneously in the same work environment requires complex scheduling and control to prevent collisions between robots, and also to prevent deadlock problems (Lee and Lee 2002). Moreover, industrial robots must be employed as effectively as possible due to the high cost of the robots. Thus, the first obvious motivation of this research is to propose a new solution to the scheduling problems in RFAC in order to overcome the above difficulties and improve the system's performance.

Due to the flexibility of RFAC and the resulting advantages, such as increased productivity with shorter cycle time, decreased labor and production costs, and increased assembling flexibility (Sect. 1.3.3), the second focus of this research has been the ability of these cells to not only assemble one type of product, but also to be adapted to assemble new products and more than one product at a time, using the same hardware, without the need for reconfiguration of the cell layout (Marian et al. 2003). This adaptation has been achieved using group technology rules when the resources of the system deal with similar parts that have the same geometrical and physical characteristics.

In real industrial situations, manufacturing systems are dynamic, due to facing unexpected events such as order cancellation, arrival of urgent orders, due date change and temporary unavailability of tools or materials. These dynamic events may cause deviations from the generated schedules, and the schedule plan may become impractical to implement when it is released to the system (Ouelhadj and Petrovic 2009). Another important motivation was to study the dynamic job shop scheduling problems of RFAC. The dynamic scheduling of RFAC to assemble more than one product is a relatively unexplored problem which is more complicated compared with static scheduling. In summary, robotic flexible assembly cell (RFAC) is a new and promising concept, which require expensive investment. In existing RFAC, scheduling decisions are the vital issues in trying to improve system utilization.

1.6 Research Gap and Scope

As mentioned earlier in this chapter, employing multi-robots in RFAC offers many advantages over RAL such as reduced surface of robotic cell, increased productivity in a shorter cycle time with lower production costs, and the unique possibility to assemble a variety of products simultaneously using the same resources. On the other hand, two robots operating simultaneously in the same workplace need a complex system to prevent collisions between them (see Appendix A). Therefore, a sophisticated scheduling system is required.

Studies on how to use the RFAC more effectively to assemble products remain limited. In the existing literature of RFAC, three vital limitations were identified. First, scheduling of RFAC in just a *single-product* assembly environment was considered. Second, scheduling of RFAC only in a *static situation* was investigated, without considering dynamic status, which reflects real world problems. Third, scheduling of RFAC only in *single-objective* optimization problems was examined in order to improve the RFAC performance. Therefore, the primary goal of this research is to cover the above research gaps by proposing new strategies that will allow decision-makers to model, simulate and optimize the scheduling of RFAC in complex environments in the most effective way. In order to achieve this goal, the scope of this thesis is divided into the following four main tasks:

- To critically review the relevant literature regarding the use of different approaches for scheduling problems in RFAC, and then highlight the major limitations that must be considered when developing a new approach (Chap. 2).
- To develop a new methodology for the static scheduling problems in RFAC based on a combination of advanced solution approaches such as simulation modelling with an artificial intelligence technique (Chap. 3). Subsequently, the developed methodology will be applied via a scenario-based case study of RFAC to demonstrate the effectiveness of this methodology (Chap. 4).
- To expand the developed methodology by considering the dynamic scheduling problems of RFAC. In this task, the important factors which influence the scheduling of RFAC will be examined. The applicability of the proposed

solution will be demonstrated via a realistic case study. Then statistical analysis tools will be applied to determine the most significant factors which affect the system performance (Chap. 5).

- To develop an optimization approach to deal with multi-objective problems for the dynamic scheduling of RFAC. In this approach, a hybrid intelligent technique will be used in order to deal with the problems which arise when the information is uncertain and ambiguous (Chap. 6). The developed approach will be verified and validated, using a realistic case study and results will be analyzed (Chap. 7).

1.7   Concluding Remarks

In this chapter, the characteristic features of the robotic flexible assembly cell (RFAC) were described and it was shown that this type of assembly system has a higher degree of dexterity and flexibility than the robotic assembly line (RAL), due to the following main reasons: the sequence of the assembly process to produce a product in RFAC is unconstrained; the design of RFAC is easy to modify and also may waste less space compared with RAL; the equipment and machines are multi-purpose. Nevertheless, the main problem of the RFAC is that more than one robot operating simultaneously in the same workplace needs a complex scheduling and control system to prevent collisions between them. To overcome the fundamental problem in RFAC, a sophisticated scheduling approach is required to guarantee higher system utilisation and ensure that the robots will move without collision.

Due to the complexity of the scheduling problems in RFAC, the procedure of the developed approach will be divided into the following four steps: (1) identifying the current research limitations for scheduling problems in RFAC; (2) developing a new optimization procedure to handle the complexity of scheduling in RFAC; (3) scheduling RFAC in a dynamic situation, which takes into consideration the significant factors influencing the system utilisation; (4) scheduling RFAC in multi-objective optimization problems in order to fully handle the uncertainty and imprecision in real world problems.

References

ABB Robotics (2013). *Operating manual: RobotStudio*. ABB Automation Technologies BA, Robotics. Vasteras, Sweden.

Browne, J., Dubois, D., Rathmill, K., Sethi, P., & Steke, K. E. (1984). Classification of flexible manufacturing systems. *Flexible Manufacturing Systems Magazine, 2*(2), 114–117.

Bukchin, J., & Tzur, M. (2000). Design of flexible assembly line to minimize equipment cost. *IIE Transactions, 32*(7), 585–598.

Burnwal, S., & Deb, S. (2013). Scheduling optimization of flexible manufacturing system using cuckoo search-based approach. *International Journal of Advanced Manufacturing Technology, 64*(5), 951–959.

Chan, F. T. S., Chan, H. K., & Lau, H. C. W. (2002). The state of the art in simulation study on FMS scheduling: A comprehensive survey. *International Journal of Advanced Manufacturing Technology, 19*(11), 830–849.

Cheng, F. S. A. (2000). A methodology for developing robotic workcell simulation models. *Winter Simulation Conference* (pp. 1265–1271). Denver, USA.

Clossen, R. J., & Malstrom, E. M. (1982). Effective capacity planning for automated factories requires workable simulation tools and responsive shop floor control. *Industrial Engineering, 15*, 73–79.

Daoud, S., Chehade, H., Yalaoui, F., & Amodeo, L. (2014). Solving a robotic assembly line balancing problem using efficient hybrid methods. *Journal of Heuristics, 20*(3), 235–259.

Delchambre, A. (1992). *Computer-aided assembly planning*. London: Chapman & Hall.

Gao, J., Sun, L., Wang, L., & Gen, M. (2009). An efficient approach for type II robotic assembly line balancing problems. *Computers & Industrial Engineering, 56*(3), 1065–1080.

Gilbert, P. R., Coupez, D., Peng, Y. M., & Delchambre, A. (1990). Scheduling of a multi-robot assembly cell. *Computer Integrated Manufacturing Systems, 3*(4), 236–245.

Groover, M. P. (2008). *Automation, production systems, and computer-integrated manufacturing*. Upper Saddle River, NJ: Pearson Education Inc.

Joseph, O. A., & Sridharan, R. (2011). Effects of routing flexibility, sequencing flexibility and scheduling decision rules on the performance of a flexible manufacturing system. *International Journal of Advanced Manufacturing Technology, 56*(1–4), 291–306.

Kazerooni, A. (1997). Scheduling flexible manufacturing systems using multiple criteria simulation (Doctor of Philosophy, University of South Australia).

Lee, J. K., & Lee, T. E. (2002). Automata-based supervisory control logic design for a multi-robot assembly cell. *International Journal of Computer Integrated Manufacturing, 15*(4), 319–334.

Lee, H. F., Srinivasan, M. M., & Yano, C. A. (2006). A framework for capacity planning and machine configuration in flexible assembly systems. *International Journal of Flexible Manufacturing Systems, 18*(4), 239–268.

Levitin, G., Rubinovitz, J., & Shnits, B. (2006). A genetic algorithm for robotic assembly line balancing. *European Journal of Operational Research, 168*(3), 811–825.

Maccarthy, B. L., & Liu, J. (1993). A new classification scheme for flexible manufacturing systems. *International Journal of Production Research, 31*(2), 299–309.

Makino, H. (1989). The comparison between robotic assembly line and robotic assembly cell. In *Proceedings of the 10th International Conference on Assembly Automation* (pp. 1–10). Canazawa, Japan.

Manivannan, S. (1993). Robotic collision avoidance in a flexible assembly cell using a dynamic knowledge base. *IEEE Transactions on Systems, Man, and Cybernetics, 23*(3), 766–782.

Marian, R. M., Kargas, A., Luong, L. H. S., & Abhary, K. (2003). A framework to planning robotic flexible assembly cells. In *32nd International Conference on Computers and Industrial Engineering* (pp. 607–615). Limerick, Ireland.

Mohamed, S. B., Petty, D. J., Harrison, D. K., & Rigby, R. (2001). A cell management system to support robotic assembly. *The International Journal of Advanced Manufacturing Technology, 18*(8), 598–604.

Nakase, N., Yamada, T., & Matsui, M. (2002). A management design approach to a simple flexible assembly system. *International Journal of Production Economics, 76*(3), 281–292.

Ouelhadj, D., & Petrovic, S. (2009). A survey of dynamic scheduling in manufacturing systems. *Journal of Scheduling, 12*(4), 417–431.

Pan, Z., Polden, J., Larkin, N., Duin, S. V., & Norrish, J. (2012). Recent progress on programming methods for industrial robots. *Robotics and Computer Integrated Manufacturing, 28*(2), 87–94.

Park, T., Lee, H., & Lee, H. (2001). FMS design model with multiple objectives using compromise programming. *International Journal of Production Research, 39*(15), 3513–3528.

Qi, L., Yin, X., Wang, H., & Tao, L. (2008). Virtual engineering: Challenges and solutions for intuitive offline programming for industrial robot. In *Proceedings of the IEEE International Conference on Robotics, Automation and Mechatronics* (pp. 12–17). Chengdu, China.

Rosati, G., Faccio, M., Carli, A., & Rossi, A. (2013). Fully flexible assembly systems (F-FAS). *Assembly Automation, 33*(2), 165–174.

Rosell, J. (2004). Assembly and task planning using Petri nets: A survey. *Journal of Engineering Manufacture, 218*(8), 987–994.

Salman, M., Ionescu, F., & Taha, R. (2009). Kinematic modeling and simulation of a scara robot by using solid dynamics and verification by MATLAB/Simulink. *European Journal of Scientific Research, 37*(3), 388–405.

Sawik, T. (1998). A lexicographic approach to bi-objective loading of a flexible assembly system. *European Journal of Operational Research, 107*(3), 656–668.

Sawik, T. (1999). *Production planning and scheduling in flexible assembly systems.* Berlin: Springer.

Sawik, T. (2004). Loading and scheduling of a flexible assembly system by mixed integer programming. *European Journal of Operational Research, 154*(1), 1–19.

Schneeweiss, C. (1995). Hierarchical structures in organizations: A conceptual framework. *European Journal of Operational Research, 86*(1), 4–31.

Stecke, K. F. (1985). Design, planning, scheduling and control problems in flexible manufacturing systems. *Annals of Operations Research, 3*(1), 3–12.

Taylan, M., & Canan, L. (2005). Mathematical modelling, simulation and experimental verification of a SCARA robot. *Simulation Modelling Practice and Theory, 13*(3), 257–271.

Udhayakumar, P., & Kumanan, S. (2010). Sequencing and scheduling of job and tool in a flexible manufacturing system using ant colony optimization algorithm. *International Journal of Advanced Manufacturing Technology, 50*(9–12), 1075–1084.

Udhayakumar, P., & Kumanan, S. (2012). Integrated scheduling of flexible manufacturing system using evolutionary algorithms. *International Journal of Advanced Manufacturing Technology, 61*(5–8), 621–635.

Yoosefelahi, A., Aminnayeri, M., Mosadegh, H., & Ardakani, H. D. (2012). Type II robotic assembly line balancing problem: An evolution strategies algorithm for a multi-objective model. *Journal of Manufacturing Systems, 31*(2), 139–151.

Zhang, D., & Qi, L. (2008). Virtual engineering: Optimal cell layout method for improving productivity for industrial robot. In *Proceedings of the IEEE International Conference on Robotics, Automation and Mechatronics* (pp. 6–11).

Zhanga, W., Freiheitb, T., & Yang, H. (2005). Dynamic scheduling in flexible assembly system based on timed Petri nets model. *Robotics & Computer Integrated Manufacturing, 21*(21), 550–558.

Chapter 2
Literature Review and Research Objectives

2.1 Introduction

The concept of robotic flexible assembly cell (RFAC) has the potential to introduce significant improvements in system performance, as highlighted in Chap. 1. To achieve these improvements, the problems of RFAC at scheduling level need to be critically addressed. To date, researchers have paid little attention to handling the scheduling problems relevant to RFAC.

The main reason for limited research on the scheduling of RFAC is because such a system requires a sophisticated *scheduling approach* not only to guarantee higher system utilisation but also to prevent collisions between robots in the shared area. Based on this restriction, the literature review in this chapter is conducted from two viewpoints: Firstly, a review of the advanced scheduling approaches that have been successfully applied to the scheduling of manufacturing systems. Secondly, a review of the existing solution approaches that have been devoted to solving scheduling of RFAC. The overall purposes of this chapter are listed as follows:

- To summarize the scheduling problems in manufacturing systems and the advanced scheduling approaches that deal with these problems;
- To review the existing approaches to scheduling problems in RFAC;
- To identify the research limitations in the existing approaches to scheduling RFAC;
- To highlight the research objectives of this thesis; and
- To introduce the research plan to achieve the research objectives.

© Springer International Publishing Switzerland 2016
K.K. Abd, *Intelligent Scheduling of Robotic Flexible Assembly Cells*,
Springer Theses, DOI 10.1007/978-3-319-26296-3_2

2.2   Scheduling Problems in Manufacturing Systems

Scheduling is a decision-making process that plays a vital role in most manufacturing industries. The scheduling function optimizes the *limited-resources* allocation to the processing of *jobs* (Pinedo 2005). Resources include machines, robots, tools, material handling equipment and materials to be processed. A job consists of a number of operations or tasks to be done in the manufacturing systems. The scheduling problems in manufacturing systems have several different aspects.

The next sub-section summarizes the three main aspects: types of scheduling problems, characteristics of scheduling problems and the solution approaches that can be used for addressing the scheduling problems in manufacturing systems.

2.2.1   Types of Scheduling Problems

Scheduling problems are theoretically categorized by a number of types. French (1982), Blazewicz et al. (1994) and Pinedo (2012) classified scheduling problems according to several criteria such as production volume, nature of production, production capacity and manufacturing systems. Each type of scheduling problem has different levels. The common types of scheduling problems with their levels are summarized in Table 2.1. In practice, the number of scheduling levels can be combined to characterize a single manufacturing environment.

Many researchers have studied the scheduling problems when the system status is either static or dynamic. These two scheduling levels are defined as follows:

- *Static scheduling* is a process that produces a fixed plan based on a set of activities when all information about the jobs is known, prior to the start of the scheduling process. The outcome of this type of scheduling cannot be changed or adapted during operating time (Jain and Elmaraghy 1997).

Table 2.1 Different types of scheduling problems (Arisha 2003)

Classification based on	Scheduling level
Production volume	High volume scheduling Intermediate volume scheduling Low volume scheduling
Nature of production	Activity scheduling Batch scheduling Network scheduling
Production capacity	Infinite capacity scheduling Finite capacity scheduling
Manufacturing systems	Flow shop scheduling Job-shop scheduling Flexible manufacturing system scheduling
State of scheduling	Static scheduling Dynamic scheduling

- *Dynamic scheduling* represents a process that creates a variable plan. Hence, dynamic scheduling is flexible, accommodating additional unexpected events such as order cancellation, arrival of urgent orders, due date changing and unavailability of tools. A dynamic scheduling plan is therefore able to respond to the market environment (Ouelhadj and Petrovic 2009a, b; Chryssolouris and Subramanian 2001).

2.2.2 Characteristics of Scheduling Problems

The characteristics of scheduling problems in manufacturing systems are specified by a set of elements. The critical elements are: decision variables, constraints, and objective functions. These elements are briefly described in the following paragraphs.

Decision variables: The decision variables consider the important elements in scheduling problems. The scheduling managers employ the decision variables to match activities and resources to finish jobs and simultaneously to optimise the system performance. The common types of decision variables (Arisha 2003) are:

- Sequencing;
- Routing;
- Timing/release; and
- Resource and activity reconfiguration.

Constraints: Most real-life scheduling problems are nondeterministic polynomial-time hard (NP-hard) and tend to have a number of constraints to generate a reliable solution. French (1982), Pinedo and Chao (1999) and Brucker (2007) listed many of the constraints that are faced in the scheduling process. The common constraints are:

- Precedence and routing constraints;
- Tooling, resources, material handling constraints;
- Industry type and automation constraints; and
- Demand pattern and capacity constraints.

Objective functions: The objective of any manufacturing company is to maximize the utilization of the resources, minimize the completion time and meet the due date of customer orders. These objectives are normally in conflict with each other and no single solution can satisfy all the objectives. Hence, the target of these companies is to strike a profitable balance among these conflicting objectives (Hopp 2001). In the scheduling area, several objective functions are used to evaluate the system's performance under different scheduling strategies. (Ramasesh 1990) categorized the objective functions into four types:

- Time-based objectives;
- Work-in-process objectives;

- Due-date-based objectives; and
- Cost-based objectives.

2.2.3 Solution Approaches

Extensive research has been done on aspects related to scheduling of manufacturing systems. Due to the complexity of scheduling problems, several solution approaches have been proposed. These approaches can be grouped into two major types: traditional and advanced solution approaches, as shown in Fig. 2.1.

Traditional approaches can be classified into two categories: analytical and heuristic approaches. These approaches can often result in near optimal solutions. However, they are applicable to only small-sized scheduling problems (Pongcharoen et al. 2002; Oduguwa et al. 2005). Traditional approaches are also inflexible, inefficient and slow to satisfy real-world scheduling problems (Shen 2002).

Advanced approaches can also be classified under two main categories, i.e. simulation approaches and artificial intelligence approaches. These approaches have also been utilised for scheduling problems in manufacturing systems. Ouelhadj and Petrovic (2009a, b) and Rajabinasab and Mansour (2011) showed that the recent studies on scheduling problems are based on simulation approaches and artificial intelligence approaches due to two reasons. First, the solutions obtained are more promising than traditional approaches. Second, advanced approaches reduce the time needed to find the solutions compared to traditional approaches. Hence, only the studies that used advanced approaches in scheduling of manufacturing systems will be reviewed here.

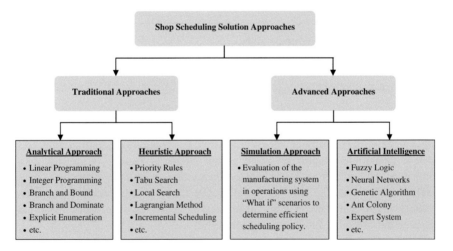

Fig. 2.1 Shop scheduling solution approaches

2.3 Review of Literature on Advanced Scheduling Approaches

There are a number of studies in the literature which look at the problems of scheduling using advanced approaches. These studies are divided into two groups as described in the next sub-section.

2.3.1 Simulation Approaches

The complexity and dynamic behavior of manufacturing systems make simulation one of the most powerful approaches for addressing scheduling problems. In the last decade, simulation-based approaches have been extensively used for scheduling problems. In these approaches, different scheduling rules are compared via simulation experiments to find the optimal solutions. For example, Mohanasundaram et al. (2002) developed efficient rules for scheduling in a dynamic assembly job shop. They compared their proposed rules with traditional scheduling rules via extensive simulation. The authors conducted the simulation experiments using two factors, namely dynamic arrival of jobs and due date changing. The simulation results indicated that the developed rules performed very well for different performance measures.

Dominic et al. (2004) combined two scheduling rules for dynamic job shop scheduling. The simulation experiments were performed using dynamic arrival of jobs and due date changing. The simulation results indicated that combined rules provided better performance. The results also showed that no single scheduling rule was effective in optimizing all performance measures.

Caprihan and Wadhwa (2005) used a simulation based approach together with Taguchi experimental design method for studying the performance of semi-automated flexible manufacturing systems under different scheduling factors. They adopted Taguchi method for conducting the experiments as well as for analyzing the simulation results.

Thiagarajan and Rajendran (2005) proposed several scheduling rules for a dynamic assembly job shop. They conducted the simulation experiments using dynamic arrival of jobs and due date changing. The results of the simulation indicated that the proposed rules were effective in optimizing multiple performance measures.

Viond and Sridharan (2008) developed five new setup oriented scheduling rules for a dynamic job shop system, and then compared the proposed rules with several common rules via experimental studies. They set up the experiments using a multi-factor experimental design: arrival of jobs, due date tightness and setup time ratio. The simulation results verified that the proposed setup oriented rules performed better than other common scheduling rules.

Ali and Wadhwa (2010) utilized simulation based approach and Taguchi method to determine the optimal combination of scheduling factors in a flexible manufacturing system (FMS). They applied Taguchi method to study the various scheduling factors and determine the important factors for improving FMS performance.

Kianfar et al. (2009) proposed new rules for dynamic flow shop scheduling. They considered dynamic arrival of jobs and the ability of acceptance and rejection of new jobs. The simulation experiments were conducted under different conditions of arrival of jobs, due date tightness and number of stages in the flow shop. The simulation results showed that the suggested new rules performed better than existing scheduling rules.

Vinod and Sridharan (2011) proposed a new methodology by investigating the interaction between scheduling rules and due date assignment methods in a dynamic job shop production system. The examined due date assignment methods were dynamic plus processing waiting time (DPPW), random work content (RWK), total work content (TWK) and dynamic total work content (DTWK). The statistical result showed a significant interaction between scheduling rules and due date assignment methods.

Caprihan et al. (2013) studied the effects of different scheduling rules on the performance of dynamic scheduling in a FMS. They considered five factors to construct a simulation experiment. These factors were information delay, routing flexibility, due date tightness, scheduling rules and dispatching rules. They used Taguchi experimental design to minimise the number of experiments as well as to analyses the simulation results. The statistical results proved that the constructed simulation model was able to demonstrate the impact of different scheduling factors on FMS performance.

2.3.2   Artificial Intelligence Approaches

Artificial intelligence approaches, such as genetic algorithms (GA), fuzzy logic (FL), ant colony optimization (ACO) and neural networks (NN) have been utilized for scheduling problems. These researchers used intelligent approaches to propose an appropriate scheduling rule, and then compared the outcome of the proposed rule with other existing scheduling rules. For instance (Tang et al. 2005) developed a rule based on NN for the dynamic scheduling of the flexible flow shop. They considered dynamic job arrivals and due date changing. The simulation results showed that the NN approach was better than all other common scheduling rules.

Vinod and Sridharan (2008) used a FL approach to propose new scheduling rules for a dynamic job shop environment. They built a discrete event simulation model to compare the proposed rules with traditional scheduling rules. The simulation experiments were conducted via due date changing and dynamic arrival of jobs. The simulation results showed that the proposed rules performed very well.

Zhou et al. (2009) proposed a new rule based on ACO for solving dynamic job shop scheduling. They examined the approach with different levels of jobs arrival, due date changing and processing time distributions. The results showed the ACO performed better than the existing scheduling rules.

Moradi (2010) integrated the flexible job-shop scheduling problems with preventive maintenance activities using GA, to reduce the probability of machine and equipment breakdowns. They employed Taguchi method to calibrate all the factors of scheduling problems and to analyze the results statistically.

Lu and Liu (2011) developed a dynamic scheduling strategy for FMS based on a FL approach. They considered arrival of jobs and due date tightness to set up the simulation experiments. The simulation results showed that the proposed strategy gave superior system performance compared to existing scheduling rules.

Nie et al. (2013) used the concept of GA to propose a rule for dynamic scheduling on the flexible job shop floor. They conducted the simulation experiments using different factors: dynamic arrival of jobs, due date changing and problem flexibility. The results indicated that the proposed rule was more efficient for complex scheduling problems.

2.3.3   Observations from the Literature Review

In the previous section, a number of studies devoted to solving dynamic scheduling problems in manufacturing systems were critically reviewed. This review showed the influential events that were considered when dealing with dynamic scheduling problems in manufacturing systems. From this reviewed literature, it can be concluded that:

- The simulation-based approaches and artificial intelligence approaches are both promising approaches for the addressing of dynamic scheduling problems. These approaches demonstrate the feasibility of studying dynamic scheduling problems and finding the optimal or near-optimal solution to maximize the utilization of manufacturing systems.
- Manufacturing systems are mostly operating in dynamic environments. These systems are often subject to factors that may cause deviations from the generated schedules, and the schedule plan may become impractical to implement when it is released to the system (Ouelhadj and Petrovic 2009a, b). Hence, the influential factors must be taken into account when solving the dynamic problems in any given production system (Jain and Elmaraghy 1997; Gholami et al. 2009).
- The Taguchi experimental design is an efficient tool for setting the number of possible experiments and for analyzing the simulation results. Taguchi method has the ability to solve scheduling problems with a greatly reduced number of experiments compared to full factorial experimental methods, in order to find the best combination of the scheduling factors to optimize the objective functions (Ali and Wadhwa 2010; Caprihan et al. 2013).

2.4 Scheduling of RFAC: A Literature Review

Much of the research on the scheduling problems has focused on flexible manufacturing systems. Researchers have paid little attention to handle the scheduling
problems in robotic flexible assembly cell (RFAC). As a result, the literature in this
area is quite limited.

This literature review can be categorized into three groups (Abd et al. 2010,
2014). The first group applied heuristic methods, while the second group investigated simulation as an approach to scheduling RFAC and the third group implemented expert systems to solve scheduling problems in RFAC.

2.4.1 Traditional Approaches

A traditional approach is an uncomplicated method to find reasonably good solutions; however it does not guarantee finding best solution. Some studies have been
dedicated to scheduling RFAC, using traditional approaches (analytical and
heuristic) as follows.

Nof and Drezner (1993) proposed robot assembly planning and scheduling
methodology relating to the allocation of assembly tasks. They formulated a
multi-robot operation as a multi-travelling salesmen problem to allow the robots to
work without collisions. They considered different configurations of multiple robot
workstations. The purpose of this study was to increase the cells' productivity and
to reduce the transportation time for robots to pick up parts and assemble them.

Lin et al. (1995) dealt with the problem of printed circuit board (PCB) assembly
when two robots are employed concurrently in the same cell. They implemented an
algorithm for simultaneous collision avoidance and scheduling operations, also to
minimize assembly cycle time and consequently enhance the throughput. The
algorithm was divided into three steps: initial insertion sequencing, balancing and
re-assignment, and avoiding collision of robots. The computational results
demonstrated the performance of the proposed algorithm.

Pelagagge et al. (1995) presented a heuristic approach to solve planning and
scheduling problems in robotic assembly cells. They focused on assembly tasks
characterization to find acceptable solutions for determining collision avoidance
and coordination problems. They divided the assembly area into two categories,
outside and inside; the latter represents critical area. The developed approach
appeared able to find acceptable solutions that guarantee high utilization of the
robots assembly.

Jiang et al. (1998) applied dynamic programming to solve the scheduling
problems for a two-robot assembly cell; these robots operated concurrently to
assemble one product. The aim of this work was to present algorithms for finding the
optimal or semi-optimal movement for each robot in the assembly cell. The main

shortcoming of this study is that, to avoid collision, just one robot has access into the assembly area and the other one stays outside the workplace at all working times.

Marian et al. (2003) proposed a framework for the planning of robotic flexible assembly cell (RFAC). This framework consists of two main modules: off-line and on-line modules. The first module was used to generate an optimal or near-optimal assembly sequence for each product, and was considered as an input module. The second module was used to determine, at every moment, the priority of assembly operations for multi-products in order to use the available resources of the RFAC. The objective was to maximize the throughput of the cell. Although they considered a simple heuristic approach to solve the planning problem of RFAC, they did not address the detailed sequencing of all assembly tasks required to assemble a product, which could result in a better solution.

2.4.2 Simulation Approaches

A simulation approach is the imitation of the operations of various real-world facilities. Many of the research studies have been devoted to developing simulation approaches for solving the decision problems in manufacturing systems, including the scheduling problems (Kianfar et al. 2009; Vinod and Sridharan 2011; Caprihan et al. 2013). However few studies, as shown below, have been done to address the scheduling problems in robotic assembly cells.

Gilbert et al. (1990) presented a procedure for the scheduling problems in a multi-robot assembly cell. The objective was to reduce the assembly time required to produce a product. They used two methods: the synchronous method, which enables on-line scheduling, and the asynchronous method, which requires off-line scheduling, but enables better assembly times. In this study, a multi-robot assembly cell was built in graphical simulation software called ROBCAD, and then simulated under different scenarios.

Hsu and Fu (1995) developed a new methodology for modelling, programming and scheduling a multi-robot assembly cell. This methodology integrated scheduling with simulation in two steps. Firstly, an AND/OR graph approach was proposed, to generate all feasible assembly sequences, and secondly, an optimal sequence of product was determined via applying a search algorithm. They built a multi-robot assembly cell by CimStation simulation software, to detect whether any collisions could happen between robots in the cell.

Basran et al. (1997) developed a flexible agent based framework for managing and operating multi-robotic assembly cells. The agents used a contract-net protocol for dynamic task allocation of assembly operations. The study divided an assembly operation into two separate stages: part fetching and part assembling. They considered a simulation approach to validate the proposed framework. The crucial shortcoming of this work is that collision avoidance between robots was completely ignored.

2.4.3   Expert System Approaches

Expert system approaches are among the artificial intelligence (AI) approaches. The basic idea of expert system approaches is to transfer the knowledge from a human to computerized systems. These systems have the ability to analyze a complex problem and recommend practicable solutions (Liao 2004). In recent years, expert systems have been extensively used to solve scheduling problems in several domains; however only three studies have been devoted to solving scheduling of RFAC.

Van Brussel (1990) proposed a knowledge-based system for scheduling flexible robotic assembly cells which incorporates task scheduling levels and real time control levels. The proposed system had the ability to create on-line scheduling by execution and monitoring of the assembly activities for a production order, from the beginning of the scheduling process to the last second the products are completed. Even though the system has the ability to schedule multi-products, no numerical examples were considered in this work.

Del Valle and Camacho (1996) proposed an expert system based approach for finding the best assembly planning and scheduling for a product in a multi-robot cell. The suggested approach was used off-line to obtain a feasible assembly plan using And/Or graph representation. The required times to change the robot tools were considered in this approach. The objective of this study was the minimization of cycle time (makespan). The results demonstrated the performance of the proposed approach with different product types.

Lee and Lee (2002) developed a strategy for scheduling and coordinating the robot tasks in a multi-robot assembly cell. They considered different types of robot tasks, namely move, tool change, pick up and assembly. They built a supervisor controlled logical system using a Petri net representation to prevent collisions between robots in a shared area. The purpose of this study was to minimize the total time of robot tasks required to assemble the final product.

2.5   Research Limitations

The literature review in Sects. 2.3 and 2.4 revealed significant limitations in the scheduling of RFAC. These limitations can be placed into three different categories: *single-product*, *static situation* and *single-objective*, as follows:

- *Scheduling of RFAC in a single-product assembly environment*: Even though the RFAC is able to assemble more than one product, the major limitation of all the studies of scheduling RFAC is that they concentrated on assembling only one type of product at a time. Therefore, there is a need for research that is related to scheduling of RFAC in a *multi-product* assembly environment.

- *Scheduling of RFAC in a static situation*: Even though the manufacturing systems in real industrial situations are essentially facing dynamic events, the second limitation of all the studies of scheduling RFAC is that they focused only on static situations without considering any dynamic events. Ouelhadj and Petrovic (2009a, b) stated that the dynamic events may cause deviations from the generated schedules, and the schedule plan may become impractical to implement when it is released to the system. Consequently, the *dynamic scheduling* of RFAC must receive further attention due to its ability to respond to the unexpected events and to provide effective solutions to real-world applications.
- *Scheduling of RFAC in single-objective optimization problems*: A number of research studies into flexible systems have attempted to optimize those scheduling problems with multi-objectives. However, nearly all the studies of scheduling RFAC are devoted only to solving single-objective optimization problems. Most of these studies used time-based objectives to evaluate the RFAC performance (e.g. Lee and Lee 2002; Marian et al. 2003). Thus, *multi-objective* functions, including time-based and due date-based objectives, need to be simultaneously optimized.

In summary, the research efforts made so far on the scheduling problems in RFAC are inadequate. Some significant issues such as *multi-product* assembly, *dynamic scheduling* and *multi-objective* optimization problems are ignored in the existing literature of RFAC, which need to be considered in the future investigations.

2.6   Research Objectives and Thesis Plan

In order to cover the research limitations in the existing literature and to find the optimal solution for the scheduling problems in RFAC, the research objectives and the plan for reaching the objectives need to be identified clearly. The research objectives and research plan of this thesis are elaborated in the following sub-sections.

2.6.1   Research Objectives

The main objective of this research is to develop a new framework for optimization of the scheduling problems in RFAC which incorporate the important issues in this area such as *multi-products, dynamic events and multi-objectives*. Due to the complexity of achieving this goal, the research objective is divided into three sub-objectives in the course of this study.

The first objective of this research is the scheduling of RFAC in a multi-product assembly environment. Consequently, this objective will overcome the major limitation in the existing literature; that it concentrates only on assembly of one type of product at a time. To achieve this objective, the following steps must be performed:

- Study the existing methodology for solving the scheduling problems in manufacturing systems.
- Develop a new methodology for multi-objective scheduling problems using intelligent technique.
- Apply the proposed methodology for solving the scheduling problems in RFAC.
- Compare and analyze the results of the developed methodology in relation to the common scheduling policy.

The second objective is the scheduling of RFAC in a dynamic situation. The aim of this objective is to examine the behavior of RFAC under dynamic events and to reflect more realistic scheduling problems. Therefore, this objective will handle the limitation of the previous studies which concentrate only on the static scheduling of RFAC. To achieve this objective, the following steps must be accomplished:

- Study the factors in the literature that have been used to solve dynamic scheduling problems and then extract the important scheduling factors that may cause deviations from the generated schedules in RFAC.
- Propose an intelligent approach for scheduling RFAC in a dynamic situation using advanced solution techniques.
- Implement the proposed approach in a scenario-based case study of RFAC, and carry out an experimental study.
- Analyze the simulation results to examine the behavior of RFAC under different scheduling factors.
- Predict the most significant scheduling factors which affect the system performance.

The third objective of this research is the optimization of dynamic scheduling for multi-objective problems. Hence, this objective will fill the gap in nearly all the relevant literature that used only one objective to evaluate the output of scheduling RFAC. This objective will also lead to improved scheduling in RFAC when two or more objective functions are simultaneously optimized. To achieve this objective, the following steps must be completed:

- Study the methods that have been devoted to solving multi-objective optimization problems in real world applications.
- Design a decision support system for solving multi-objective optimization problems.
- Develop a hybrid approach, using a combination of advanced solution methods.
- Demonstrate the applicability of the developed approach via a realistic case study.

- Analyze the results of the developed approach to predict the optimal combinations of scheduling factors to optimize the objective functions.
- Verify and validate the results obtained, through a sensitivity analysis and confirmation test.

2.6.2 Research Plan and Thesis Structure

In order to meet the research objectives described in the previous sub-section, a comprehensive research plan is required for finding an optimal solution to the scheduling problems in RFAC. In this study, the research plan is divided into three stages, which are stated below along with the procedures taken to achieve them. An outline of the chapters comprising this thesis is also presented in this section.

Stage 1: *New methodology for solving scheduling problems in RFAC*
In the first stage, a new methodology for scheduling RFAC in a multi-product assembly environment will be developed. In this methodology, a fuzzy-based mathematical model will be combined with simulation modelling to generate an efficient schedule for assembling more than one product at a time.

The developed methodology will be divided into three modules: *pre-processing*, *scheduling* and *simulation*. In the *pre-processing* module, the parameters, objective functions, constraints and decision variables of modelling the scheduling problems in the RFAC will be defined. In the *scheduling* module, the schedule for assembling multi-products will be generated via a new scheduling rule, namely fuzzy sequencing rule (FSR). In the *simulation* module, a computer simulation model of the RFAC will be built using simulation software, and then simulated under different scenarios. The detailed construction of this new methodology will be presented in Chap. 3.

The verification of the methodology's usefulness and effectiveness will be performed via an extensive example application and then the results obtained by the methodology will be analyzed and discussed. The application and results analysis will be explained in detail in Chap. 4.

Stage 2: *Development of an approach for scheduling RFAC in a dynamic situation*
In the second stage, a new approach for the dynamic scheduling of RFAC will be proposed. The developed approach will be divided into four phases: *preparation*, *Taguchi method*, *simulation modelling* and *statistical analysis*.

In the *preparation* phase, the scheduling problems, objectives, scheduling factors and the number of levels for each factor will be identified, and each scheduling factor will be assigned three levels, because the influence of these factors may vary nonlinearly. In the *Taguchi method* phase, the number of possible experiments will be determined on the basis of the number of factors and their levels. The benefits of using Taguchi optimization method are to reduce the minimum number of experiments required for scheduling RFAC and to gain better understanding of the impact of the scheduling factors on system performance. In the *simulation modelling* phase, the RFAC will be built as a computer model to evaluate their performance. In the *statistical analysis* phase, analysis of mean (ANOM), relative percentage deviation (RPD) and analysis of variance (ANOVA) will be conducted. The ANOM and RPD will be used to determine the optimal level of each scheduling factor based on S/N ratio results; the ANOVA will be applied to determine the most significant scheduling factors in terms of their contribution to the objective functions. The detailed framework for developing the new approach and its application to a realistic case study of RFAC will be presented in Chap. 5.

Stage 3: *Development of an optimization approach with multi-objective problems*
In the last stage of this thesis, an optimization approach for multi-objective problems in dynamic scheduling of RFAC will be developed. This approach will be based on hybrid techniques: the fuzzy decision support system (FDSS) and the fuzzy analytic hierarchy process with the fuzzy technique for order preference by similarity to ideal solution (FAHP-FTOPSIS). In FDSS, a computer-based system will be designed using the Matlab fuzzy logic toolbox. In FAHP-FTOPSIS, the FAHP and FTOPSIS will be integrated to optimize the dynamic scheduling in RFAC under different objective-functions. The developed approach will be divided into three phases: *problem description*; *application of fuzzy MCDM*; and *analysis of the results*.

In the *problem description* phase, the hierarchical structure of any MCDM problem will be defined. This hierarchy is constructed based on the overall goal, criteria and sub-criteria (objective functions), and the decision alternatives (feasible solutions). In the *application of the fuzzy MCDM* phase, the FDSS and FAHP-FTOPSIS will be applied to evaluate the feasible solutions and make a final decision. In the analysis of results phase, the outcomes of FDSS and FAHP-FTOPSIS will be verified and validated. The comprehensive detail of the above three phases that are followed to develop the hybrid fuzzy MCDM approach will be presented in Chap. 6.

A realistic case study will be employed to test the efficiency of the hybrid optimisation approach. The analysis tools such as sensitivity analysis and confirmation test will be performed to verify and validate the results obtained. The case study and analysis of the results will be presented in Chap. 7.

The summary of the research presented in this thesis, the contributions of the research and possible recommendations for future research will be presented in Chap. 8.

The three sub-objectives of this research and the outline structure of the thesis are shown in Fig. 2.2.

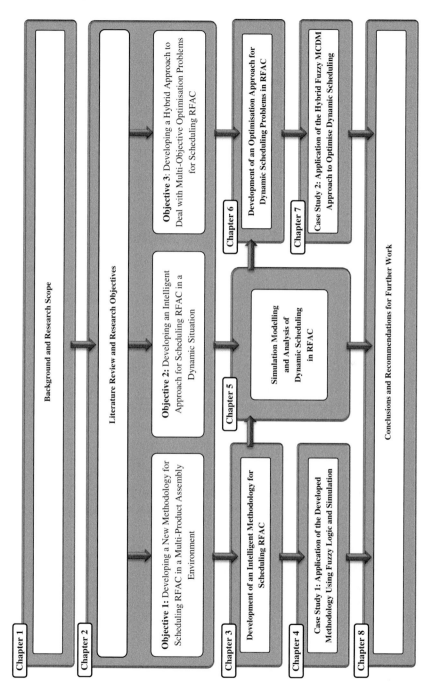

Fig. 2.2 Outline of thesis structure

2.7 Concluding Remarks

In this chapter, the literature on scheduling problems in manufacturing systems was reviewed. This literature review revealed that there has been relatively little work on the scheduling of robotic flexible assembly cell (RFAC), even though overall scheduling problems of flexible manufacturing systems (FMS) have attracted significant attention. The reason for this is because the RFAC requires a complex scheduling system to prevent collisions between robots. Consequently, the decision making in scheduling RFAC is more difficult compared with decision making the in FMS.

In the existing literature of RFAC, three important limitations were identified. First, scheduling of RFAC in a *single-product* assembly environment was considered. Second, scheduling of RFAC only in a *static situation* was investigated without considering dynamic status, which reflects the real world problems. Third, scheduling of RFAC just in *single-objective* optimization problems was examined to improve the RFAC performance. To overcome these limitations, an efficient and novel approach for scheduling RFAC is required to deal with various issues: *multi-product* assembly environment, *dynamic status* and *multi-objective* optimization problems.

From the literature survey in this chapter, it can be seen that there are two approaches to addressing the scheduling problems in manufacturing systems. First, researchers are working towards developing *traditional approaches* to easily find feasible solutions without optimization. Second, researchers are working towards developing *advanced approaches* (e.g. simulation, artificial intelligence) to accurately find optimal or near-optimal solutions. Thus, the main objective of this research is to develop an advanced approach for tackling the scheduling problems in RFAC. In the course of the research, the proposed objective is divided into three sub-objectives: the first objective is developing a new methodology for robust scheduling of RFAC in a multi-product assembly environment (Chaps. 3 and 4); the second objective is developing an intelligent approach for scheduling RFAC in a dynamic situation (Chap. 5); the third objective is developing a hybrid approach to deal with multi-objective optimization problems for scheduling RFAC (Chaps. 6 and 7).

References

Abd, K., Abhary, K., & Marian, R. (2010). A scheduling framework for robotic flexible assembly cells. *AIJSTPME-Asian International Journal of Science and Technology in Production and Manufacturing Engineering, 4*(1), 30–37.

Abd, K., Abhary, K., & Marian, R. (2014). Simulation modelling and analysis of scheduling in robotic flexible assembly cells using Taguchi method. *International Journal of Production Research, 52*(12), 2654–2666.

Ali, M., & Wadhwa, S. (2010). The effect of routing flexibility on a flexible system of integrated manufacturing. *International Journal of Production Research, 48*(19), 5691–5709.

Arisha, A. (2003). *Intelligent shop scheduling for semiconductor manufacturing.* Doctor of Philosophy: Dublin City University.

Basran, J. S., Petriu, E. M., & Petriu, D. C. (1997), Flexible agent-based robotic assembly cell. In *Proceedings of the IEEE International Conference on Robotics and Automation* (pp. 3461–3466). Albuquerque, New Mexico.

Blazewicz, J., Ecker, K. H., Schmidt, G., & Weglarz, J. (1994). *Scheduling in computer and manufacturing systems*. Berlin: Springer Verlag.

Brucker, P. (2007). *Scheduling algorithms*. Berlin: Springer.

Caprihan, R., & Wadhwa, S. (2005). Scheduling of FMSs with information delays: A simulation study. *International Journal of Flexible Manufacturing Systems, 17*(1), 39–65.

Caprihan, R., Kumar, A., & Stecke, K. E. (2013). Evaluation of the impact of information delays on flexible manufacturing systems performance in dynamic scheduling environments. *The International Journal of Advanced Manufacturing Technology, 67*(1), 311–338.

Chryssolouris, G., & Subramanian, E. (2001). Dynamic scheduling of manufacturing job shops using genetic algorithm. *Journal of Intelligent Manufacturing, 12*(3), 281–293.

Del Valle, C., & Camacho, E. F. (1996). Automatic assembly task assignment for a multirobot environment. *Control engineering practice, 4*(7), 915–921.

Dominic, P. D. D., Kaliyamoorthy, S., & Kumar, S. M. (2004). Efficient Dispatching Rules for Dynamic Job Shop Scheduling. *International Journal Advance Manufacturing Technology, 24*(1), 70–75.

French, S. (1982). *Sequencing and scheduling: An introduction to the mathematics of the job-shop*. New York.

Gholami, M., Zandieh, M., & Alem-Tabriz, A. (2009). Scheduling hybrid flow shop with sequence-dependent setup times and machines with random breakdowns. *International Journal of Advanced Manufacturing Technology, 42*(1), 189–201.

Gilbert, P. R., Coupez, D., Peng, Y. M., & Delchambre, A. (1990). Scheduling of a multi-robot assembly cell. *Computer Integrated Manufacturing Systems, 3*(4), 236–245.

Hopp, W. J. (2001). *Factory physics: Foundations of manufacturing management*. New York: McGraw-Hill.

Hsu, H. H., & Fu, L. C. (1995). Fully automated robotic assembly cell: Scheduling and Simulation. In *Proceedings of the IEEE International Conference on Robotic and Automation* (pp. 208–214). Nagoya, Japan.

Jain, A. K., & Elmaraghy, H. A. (1997). Production scheduling rescheduling in flexible manufacturing. *International Journal of Production Research, 35*(1), 281–309.

Jiang, K., Seneviratne, L. D., & Earles, S. W. E. (1998). Scheduling and compression for a multiple robot assembly workcell. *Production Planning and Control: The Management of Operations, 9*(2), 143–154.

Kianfar, K., Fatemi Ghomi, S. M. T., & Karimi, B. (2009). New dispatching rules to minimize rejection and tardiness costs in a dynamic flexible flow shop. *International Journal of Advanced Manufacturing Technology, 45*(7), 759–771.

Lee, J. K., & Lee, T. E. (2002). Automata-based supervisory control logic design for a multi-robot assembly cell. *International Journal of Computer Integrated Manufacturing, 15*(4), 319–334.

Liao, S. H. (2004). Expert system methodologies and applications-a decade review from 1995 to 2004. *Expert Systems with Applications, 28*(1), 93–103.

Lin, H. C., Egbelu, P. J., & Wu, C. T. (1995). A two-robot printed circuit board assembly system. *International Journal of Computer Integrated Manufacturing, 8*(1), 21–31.

Lu, M. S., & Liu, Y. J. (2011). Dynamic dispatching for a flexible manufacturing system based on fuzzy logic. *The International Journal of Advanced Manufacturing Technology, 54*(9–12), 1057–1065.

Marian, R. M., Kargas, A., Luong, L. H. S., & Abhary, K. (2003). A framework to planning robotic flexible assembly cells. In *32nd International Conference on Computers and Industrial Engineering* (pp. 607–615). Limerick, Ireland.

Mohanasundaram, K. M., Natarajan, K., Viswanathkumar, G., Radhakrishnana, P., & Rajendran, C. (2002). Scheduling rules for dynamic shops that manufacture multi-level jobs. *Computers & Industrial Engineering, 44*(1), 119–131.

Moradi, E., Ghomi, S. M. T. F., & Zandieh, M. (2010). An efficient architecture for scheduling flexible job-shop with machine availability constraints. *The International Journal of Advanced Manufacturing Technology, 51*(1–4), 325–339.

Nie, L., Gao, L., Li, P., & Li, X. (2013). A GEP-based reactive scheduling policies constructing approach for dynamic flexible job shop scheduling problem with job release dates. *Journal of Intelligent Manufacturing, 4*(4), 763–774.

Nof, S. Y., & Drezner, Z. (1993). The Multiple-Robot Assembly Plan Problem. *Journal of Intelligent and Robotic Systems, 7*(1), 57–71.

Oduguwa, V., Tiwari, A., & Roy, R. (2005). Evolutionary computing in manufacturing industry: An overview of recent applications. *Applied Soft Computing, 5*(3), 281–299.

Ouelhadj, D., & Petrovic, S. (2009a). A survey of dynamic scheduling in manufacturing systems. *Journal of Scheduling, 12*(4), 417–431.

Ouelhadj, D., & Petrovic, S. (2009b). A survey of dynamic scheduling in manufacturing systems. *Journal of Scheduling, 12*(4), 417–431.

Pelagagge, P. M., Cardarelli, G., & Palumbo, M. (1995). Design criteria for cooperating robots assembly cells. *Journal of Manufacturing Systems, 14*(4), 219–229.

Pinedo, M., & Chao, X. (1999). *Operations scheduling with applications in manufacturing and services*. New York: McGraw-Hill.

Pinedo, M. (2005). *Planning and scheduling in manufacturing and services*. New York: Springer.

Pinedo, M. (2012). *Scheduling: Theory, algorithms, and systems*. New York: Springer.

Pongcharoen, P., Hicks, C., Braiden, P. M., & Stewardson, D. J. (2002). Determining optimum genetic algorithm parameters for scheduling the manufacturing and assembly of complex products. *International Journal of Production Economics, 78*(3), 311–322.

Rajabinasab, A., & Mansour, S. (2011). Dynamic flexible job shop scheduling with alternative process plans: an agent-based approach. *International Journal of Advanced Manufacturing Technology, 54*(9–12), 1091–1107.

Ramasesh, R. (1990). Dynamic job shop scheduling: A survey of simulation research. *OMEGA: International Journal of Management Science, 18*(1), 43–57.

Shen, W. (2002). Distributed manufacturing scheduling using intelligent agents. *IEEE Intelligent Systems, 17*(1), 88–94.

Tang, L., Liu, W., & Liu, J. (2005). A neural network model and algorithm for the hybrid flow shop scheduling problem in a dynamic environment. *Journal of Intelligent Manufacturing, 16*(3), 361–370.

Thiagarajan, S., & Rajendran, C. (2005). Scheduling in dynamic assembly job-shops to minimize the sum of weighted earliness, weighted tardiness and weighted flowtime of jobs. *Computers & Industrial Engineering, 49*(4), 463–503.

Van Brussel, H., Cottrez, F., & Valckenaers, P. (1990). ASESFAC: A scheduling expert system for flexible assembly cells. In *International Conference on Robotics & Automation* (pp. 1950–1955). Cincinnati, Ohio.

Vinod, V., & Sridharan, R. (2008). Development and analysis of fuzzy priority rules for scheduling a dynamic job shop production system. In *Proceedings of the IEEE International conference on Industrial Engineering and Engineering Management* (pp. 1418–1422).

Vinod, V., & Sridharan, R. (2011). Simulation modeling and analysis of due-date assignment methods and scheduling decision rules in a dynamic job shop production system. *International Journal of Production Economics, 129*(1), 127–146.

Viond, V., & Sridharan, R. (2008). Dynamic job-shop scheduling with sequence-dependent setup times: Simulation modeling and analysis. *International Journal of Advanced Manufacturing Technology, 36*(3–4), 355–372.

Zhou, R., Nee, A. Y. C., & Lee, H. P. (2009). Performance of an ant colony optimisation algorithm in dynamic job shop scheduling problems. *International Journal of Production Research, 47*(11), 2903–2920.

Chapter 3
Development of an Intelligent Methodology for Scheduling RFAC

3.1 Introduction

Production scheduling of advanced manufacturing systems has attracted significant attention by both researchers and industrial practitioners in recent years. Due to the complexity of these systems, the generation of production schedules requires an intelligent technique. Many artificial intelligence techniques such as fuzzy logic (FL), genetic algorithms (GA) and neural networks (NN) have been successfully applied to the scheduling of advanced manufacturing systems (Chan and Chan 2004; Xing et al. 2010). As mentioned in Chap. 1, RFAC is considered to be one class of such systems. Consequently, the objective of this part of the research is to propose a new intelligent methodology of scheduling RFAC in a multi-product assembly environment, using FL. To achieve this, the following two main steps have to be accomplished.

- Review of the relevant literature for the use of fuzzy logic approaches for scheduling problems in both conventional and flexible manufacturing systems, and then extraction of the key points to be considered when developing a conceptual methodology for scheduling RFAC.
- Development of a new methodology for multi-objective scheduling problems in RFAC. The proposed methodology is based on combining a fuzzy-based mathematical model with simulation software called SIMPROCESS.

3.2 Application of Fuzzy Logic to Scheduling Problems

In manufacturing systems, since the scheduling problems are nondeterministic polynomial-time hard (NP-hard), an efficient approach is required to get best results (Buil et al. 2010; Sridhar et al. 2010). Recently, a fuzzy logic approach has been widely applied to the scheduling problems for both conventional and flexible

© Springer International Publishing Switzerland 2016
K.K. Abd, *Intelligent Scheduling of Robotic Flexible Assembly Cells*,
Springer Theses, DOI 10.1007/978-3-319-26296-3_3

manufacturing systems (Subramaniam et al. 2000; Vidyarthi and Tiwari 2001; Domingos and Politano 2003; Bilkay et al. 2004; Canbolat and Gundogar 2004; Kumar et al. 2004; Srinoi et al. 2006; Restrepo and Balakrishnan 2008; Srinoi et al. 2008; Mahdavi et al. 2009).

In this section, a brief overview of relevant studies and how these studies on the application of fuzzy logic have been applied to solving scheduling problems is presented. This literature review will also provide the necessary key points for the development of a conceptual methodology for the scheduling of RFAC.

Subramaniam et al. (2000) developed a scheduling method, named the fuzzy scheduler, based on fuzzy logic which was used in the job shop environment to evaluate several candidate scheduling rules and then select the most convenient rule. In this study, the fuzzy scheduler was applied to different industrial job shop problems and it was demonstrated that the fuzzy scheduler performs better than other common scheduling rules. The results also showed that the fuzzy scheduler is as easy to apply as the scheduling rules.

Vidyarthi and Tiwari (2001) presented a fuzzy-based methodology for machine loading in a FMS. They considered processing time, batch size and optional operation processing time as three fuzzy input variables. The first step in this methodology was to evaluate the overall contribution of the job variables, using fuzzy membership functions, and then to determine the job sequencing as the output fuzzy variable. The objective was to maximize the throughput and minimize the system unbalance.

Domingos and Politano (2003) proposed a procedure based on fuzzy logic for scheduling a FMS. In this study, processing time, due date and work load were the input fuzzy variables, and the part priority was the output fuzzy variable. Three common objective functions, namely average tardiness, percentage of tardy jobs and average flow time, were to be minimized. They compared their procedure with common scheduling rules via simulation software. The simulation results indicated that the proposed procedure improved the performance of the FMS under multi-objectives.

Kumar et al. (2004) developed a fuzzy based algorithm to solve the scheduling problems of a FMS. They applied fuzzy membership functions to evaluate the overall contribution of each job type to the objectives according to the selected attributes, and then determine the job sequencing. They used processing time, batch size and required tool slots as the main attributes. Two objective functions, maximizing of throughput and minimizing of system imbalance, were considered in this study. The computational results showed that the developed algorithm gives better solutions than those obtained by heuristic approaches.

Bilkay et al. (2004) proposed a fuzzy logic-based decision-making algorithm for generating the sequence of part types to be processed. They generated a sequence based on priority value of the part types, from higher priority to lower priority value. The algorithm prioritized part types according to four input variables: processing time, batch size, due date and the required tool slots. The results showed that fuzzy logic improves the system efficiency and is suited for scheduling problems that have multiple conflict objectives.

Canbolat and Gundogar (2004) applied a fuzzy logic approach to solve a multi criteria scheduling problem for a job shop environment. The suggested approach combined three scheduling rules in a new rule named fuzzy priority rule (FPR). The new rule was compared with other traditional scheduling rules such as SPT, EDD, CR, etc., using a simulation program. The simulation results showed the superiority of the FPR over traditional rules in mean flow time and mean tardiness.

Srinoi et al. (2006) developed a new approach based on fuzzy logic to generate a scheduling model for solving the resource allocation problem in flexible manufacturing systems. They defined four fuzzy input variables of the model: processing time, due date, setup time and machine priority; the output variable of the model was the job priority. They conducted several experiments to prove the effectiveness of the developed approach. The experimental results indicated that the fuzzy logic approach is a powerful technique for scheduling problems in FMS, based on multi criteria objectives.

Restrepo and Balakrishnan (2008) proposed a fuzzy-based methodology to solve multi-objective scheduling problems for robotic flexible manufacturing cells. The proposed methodology compared two scheduling rules namely SPT and EDD. Simulation results showed that the existing scheduling rules were satisfied only when considering single objectives such as maximizing the throughput time or minimizing the tardiness cost. The results indicated also that the fuzzy-based methodology is able to give a better performance than SPT and EDD.

Srinoi et al. (2008) developed a fuzzy-based mathematical model to deal with scheduling in FMS, based on multi-performance measures. They used processing time, machine priority, due date and setup time as input fuzzy variables, while the job priorities were the output variable. The simulation results pointed out the superiority of the suggested model in most performance measures.

Mahdavi et al. (2009) presented a fuzzy approach to solve the scheduling problems of a FMS. They defined four fuzzy input variables: processing time, workload, setup time and travelling time. In this study, the output fuzzy variable was the optimal route selection to satisfy multi-conflicting objectives. They used MATLAB fuzzy logic toolbox to determine the route selection. The numerical results showed that the presented approach is easily applicable to finding the optimal flexible routing in FMS.

Based on the previous studies, five key points can be extracted:

- Most of the above studies showed that the use of fuzzy logic and simulation tools is suitable to optimize the scheduling problems for both conventional and flexible manufacturing systems.
- The scheduling problems may be divided into three main sub-problems: part type selection, machine loading and resource allocation. Most of the studies dealt with part type selection problems.
- The majority of studies took into account the processing time; due date and batch size were the main fuzzy criteria.

- Two decision types, parts routing and parts sequencing, can be identified in the above studies. Most of the studies focused on the parts sequencing decision.
- Nearly all the studies reviewed in this section employed more than two performance measures to evaluate the quality of the schedules.

This chapter attempts to use the key points mentioned above to develop an intelligent methodology for scheduling of RFAC. Therefore, the proposed methodology will include the following: the technique that will be used is fuzzy logic in combination with a simulation tool; the scheduling problem is product type selection; the fuzzy criteria are processing time, due date and batch size; the decision type is products sequencing; and the scheduling output is evaluated using multi performance measures. The next section will describe the proposed methodology in more detail.

3.3 Proposed Methodology for Scheduling of RFAC

The main purpose of this chapter is to explain the development of a methodology that allows the user to model the scheduling of RFAC more effectively (Abd et al. 2013a, b, c). The scheduling of the RFAC requires finding a way which determines how best to use cell resources to assemble multi-products. Let us consider an assembly cell in which a set of tasks is performed using a set of resources to concurrently assemble multi-products.

- Tasks represent any physical activities that are carried out by utilizing resources. The tasks can be categorized into four types: move, tool-change, pick-up and assembly.
- Multi-products of the same family group usually involve similar operations; however, there will be some differences in the assembly operations and the operational sequences between these products.

In this section, the architecture of the proposed methodology is described (Fig. 3.1). This methodology has three major modules as explained below.

- Pre-processing module: this module helps to define the components of the scheduling problem model. For example, this module determines the system's inputs/output, identifies the objectives and describes the characteristics of RFAC.
- Scheduling module: this module is the core of the proposed methodology, which allows the user to generate the schedule for assembling multi products.
- Simulation module: this enables the user to build the RFAC as a computer model, and then simulates the model under different scenarios, depending on the outcome of the scheduling module.

Fig. 3.1 Architecture of proposed methodology for RFAC scheduling

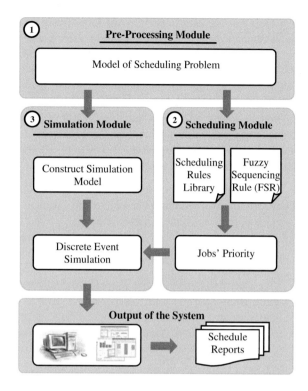

3.3.1 Pre-processing Module

The aim of the pre-processing module is to describe all the required components of the scheduling problem model in the RFAC. These components are shown in Fig. 3.2 and described below.

Parameters: The required parameters for the scheduling process are *system structure parameters* and *jobs parameters*. The *system structure parameters* depend on the configuration of the system. In other words, they reflect the physical characteristics of the system. For example, RFAC generally consist of main resources

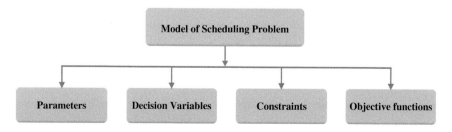

Fig. 3.2 Model of scheduling problem

and tools that are used to perform the jobs. These resources are: robots for fetching the assembled parts and placing them at a number of assembly stations (AS_1, AS_2, ..., AS_n); parts feeder (PF) for supplying parts to the cell; gripper changing station (GC); input conveyor (IC) for supplying the base parts; and output conveyor (OC) for conveying out a final product when assembly processes are completed (Marian et al. 2003; Abd et al. 2011). *Jobs parameters* represent inputs data for a system: in other words, input variables that have fixed values. In this methodology, processing time, batch size and due date are selected as the important input variables in the scheduling problems (Bilkay et al. 2004). Also, the number of required stations is considered as another variable.

- Processing time: this input variable represents the summation time of all required tasks needed to complete the product. These tasks are pickup and release, parts movement and assembly.
- Batch size: any flexible system can process different jobs. Each job is processed in a different amount called a batch size, which depends on the customer requirements.
- Due date: this input variable denotes the deadline of production for each job. In other words, the job must be completed prior to the time required by the customer; otherwise the company might face a penalty for late completion time.
- Number of required stations: the last input variable gives high priority to the product which requires the greater number of stations.

Decision Variables: In this research, the decision variable is represented by the job priority, illustrating the priority status of a product to be selected for the next assembly operation in RFAC. The scheduling module section will explain how the job priority using scheduling rules is determined.

Constraints: Constraints define the feasibility of a schedule. To generate a reliable solution to practical problems, a set of constraints must be satisfied. In this research, the RFAC scheduling problem is subject to three resource constraints: tooling resource constraints, robot movement constraints and robot access constraints (Abd et al. 2013b).

- To fetch and assemble, the hand of each robot should be equipped with the right tool; however, a specific tool may not be available for the two robots simultaneously, due to the restricted number of available tools. These are tooling resource constraints.
- Robot arms cannot move from one place to another directly. The reason for this is to avoid collisions with the other robot arms. This is achieved by assigning control points in the cell. Control points $\{C_1, C_2, ..., C_4\}$ are set to simplify path planning and avoid collisions. For example, R1 cannot move from S_5 to S_6 directly, rather via control point C_2. These requirements are called robot movement constraints, as shown in Fig. 3.3.
- To prevent collisions between robots in a shared area, more than one robot cannot access the same resource at the same time. For instance, just one robot, R_1 or R_2, can access transfer table (S_4) or tool magazine (S_5) or assembly station

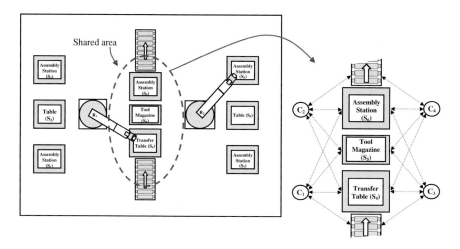

Fig. 3.3 Robot move and access constraints

(S_6) or the conveyors IN and OUT. These requirements are named robot access constraints, as shown in Fig. 3.3.

Objective functions: The objective function is a quantity to be minimized or maximized in any optimization problem. Examples of objective functions include makespan, system utilization, lateness/tardiness, production cost. In the scheduling area, several objective functions are used to evaluate the system's performance. (Ramasesh 1990) categorized the objective functions into four types: time-based objectives, work-in-process objectives, due-date-based objectives and cost-based objectives.

In this research, five objective functions, namely makespan, percentage of robots idle time, total tardiness, maximum tardiness and percentage of tardy jobs, are to be minimized, to evaluate the RFAC performance under different scheduling policies. These objectives are classified into two categories: time based objectives and due date based objectives. The makespan and percentage of robots idle time are in the first category while total tardiness, maximum tardiness and percentage of tardy jobs fall into the second category. The following notations are used to formulate the mathematical expressions of the objectives.

P	Products index $(P = 1, 2, \ldots, i)$
Q	Part index $(Q = 1, \ldots, j)$
R	Robots index $(R = 1, 2, \ldots, k)$
S	Resource index $(S = 1, \ldots, l)$
OP	Assembly operation index $(OP = op_{1i}, op_{2i}, \ldots, op_{mi})$ of product i
T_{mi}	Time of assembly operation m of product i

(continued)

(continued)

$T_{(s \to l)i}$	Time taken by robot to travel between two resources ($s \to l$), to assemble product i
T_{ji}	Time of tool change to transfer/assemble component j of product i
D_i	Due date of product i
N_i	Batch size of product i
C_i	Completion time of product i
U_i	Indicator for whether product i is tardy or not

Makespan (C_{max}): The makespan is the maximum completion time of the last job processed by robots. The minimization of this objective results in an efficient utilization of system resources (Danping and Lee 2010). The makespan can be represented as:

$$C_{max} = \max_{1 \le i \le p} (C_i) \quad \forall R \tag{3.1}$$

Percentage of robots idle time ($\%I_T$): The robots idle time is the time the robots must sit idle (Desal 1997): in other words, the waiting time of robots before the start of any actions such as movement, tool-change, pick up and assembly. The percentage of robots idle time can be calculated using the following formula:

$$\%I_T = \left(1 - \frac{\sum_{OP=1}^{m} T_{mi} + \sum_{S=1}^{l} T_{(s \to l)i} + \sum_{Q=1}^{j} T_{ji}}{C_{max}} \right) \times 100 \quad \forall i \tag{3.2}$$

Total tardiness (TD): The total tardiness is the sum of the tardiness of all jobs (Restrepo and Balakrishnan 2008; Berrichi and Yalaoui 2013). Minimization of total tardiness aims to find schedules that satisfy the customers' due dates. Total tardiness can be represented as:

$$TD = \sum_{p=1}^{i} [C_i - D_i, 0] \tag{3.3}$$

Maximum tardiness (MaxTD): The maximum tardiness is the largest difference between the completion time and the due date committed to for all jobs (Tavakkoli-Moghaddam et al. 2005; Baptiste and Schieber 2003). The maximum tardiness can be represented using the following formula:

$$MaxTD = \left(\max_{1 \le i \le p} [C_i - D_i] \right) \tag{3.4}$$

Percentage of tardy jobs ($\%N_U$): The percentage of tardy jobs is the sum of overdue jobs divided by the number of jobs (Domingos and Politano 2003;

Jayamohan and Rajendran 2000). The percentage of tardy jobs can be expressed using the following formula:

$$\%N_U = \frac{\sum_{P=1}^{i} U_i}{\sum_{P=1}^{i} N_i} \times 100, \quad U_i = \begin{cases} 1, & if\ C_i > D_i \\ 0, & otherwise \end{cases} \tag{3.5}$$

3.3.2 Scheduling Module

In scheduling RFAC, when a robot becomes free and more than one job is waiting for processing, the jobs will be scheduled, from the highest priority to the lowest priority. This can be done using scheduling rules. These rules are used to generate the sequence of job flow to the system. In this research, each product is considered as an independent job. The algorithm of the scheduling module is depicted in Fig. 3.4.

In the proposed methodology, the scheduling module contains two types of rules which allow the decision maker to determine the job sequencing.

The first type is the rules that have been commonly used in scheduling for solving scheduling problems. The following is a list of the common scheduling rules used herein.

- Short Processing Time (SPT): select the job with minimum processing time first.
- Long Processing Time (LPT): select the job with maximum processing time first.
- Random (RAND): jobs are sequenced randomly.
- Earlier Due Date (EDD): jobs are sequenced according to their due dates.
- Critical Ratio (CR): select the job with minimum critical ratio first.
- Minimize Slack Time (MST): jobs are sequenced according to their urgency.

The second type is a rule developed for scheduling RFAC in a multi-product assembly environment, called a fuzzy sequencing rule (FSR) which is constructed by combining all the input variables using fuzzy logic technique. In this research, the job sequence determination is carried out by evaluating the normalization of each job variable such as processing time, batch size, due date and number of required stations. The normalization of the four inputs to the system can be defined, using the following notations:

μ_T^i	Normalization of the processing time T of product i
μ_D^i	Normalization of the due date D of product i
μ_N^i	Normalization of the batch size N of product i
μ_S^i	Normalization of the number of required stations S for product i

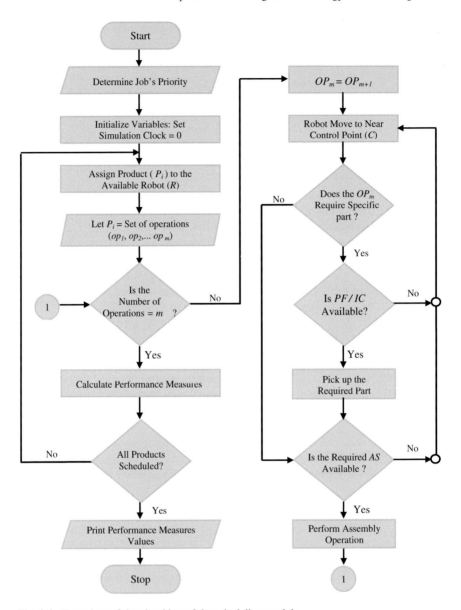

Fig. 3.4 Flow chart of the algorithm of the scheduling module

μ_T^i: The normalisation of the total processing time of product i is defined as the ratio of the difference between the total processing time of product i and its minimum total processing time to the difference between the maximum and minimum total processing times of the same product, as shown in Eq. 3.6.

$$\mu_T^i = \frac{[(T_i) - Min(T_i)]}{[Max(T_i) - Min(T_i)]}, \quad 0 \le \mu_T^i \le 1 \tag{3.6}$$

According to the Eq. 3.6, it can be seen that the product with minimum processing time has a normalized value of 0 and the product with maximum processing time has a normalized value equal to 1.

μ_D^i: The normalisation of the due date of product i is defined as the ratio of the difference between the due date of product i and its minimum due date to the difference between the maximum and minimum due dates, as expressed in Eq. 3.7.

$$\mu_D^i = \frac{[(D_i) - Min(D_i)]}{[Max\ (D_i) - Min\ (D_i)]}, \quad 0 \le \mu_D^i \le 1 \tag{3.7}$$

From this it can be concluded that the product with minimum due date has a normalized value of 0 and the product with maximum due date has a normalized value of 1.

μ_N^i: The normalization of the batch size of product i is defined as the ratio of the difference between the batch size of product i and its minimum batch size to the difference between the maximum and minimum batch size of product i, as shown in Eq. 3.8.

$$\mu_N^i = \frac{[(N_i) - Min(N_i)]}{[Max(N_i) - Min(N_i)]}, \quad 0 \le \mu_N^i \le 1 \tag{3.8}$$

Again, Eq. 3.8 shows that the product with minimum due date has a normalized value equal to 0 and that with maximum due date has a normalized value of 1.

μ_S^i: The normalization of the number of required stations for product i is defined as the ratio of the difference between the maximum number of required stations of product i and its number of required stations to the difference between the maximum and minimum numbers of required stations for product i, as expressed in Eq. 3.9.

$$\mu_S^i = \frac{[Max(S_i) - (S_i)]}{[Max(S_i) - Min(S_i)]}, \quad 0 \le \mu_S^i \le 1 \tag{3.9}$$

According to Eq. 3.9, the product with the minimum number of required stations has a normalized value of 1 and the product with the maximum number of required stations has a normalized value of 0.

All these normalizations are combined to determine which product must be assembled first. The products with low μ_T^i, early μ_D^i, low μ_N^i and high μ_S^i will take high priority. Now a mathematical model is developed to calculate the jobs priority, using fuzzy logic. Section 3.4 will explain the implementation of this fuzzy-based mathematical model. As mentioned in Chap. 2, FLS consist of four components:

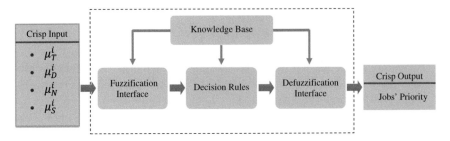

Fig. 3.5 Fuzzy logic system configuration for job selection

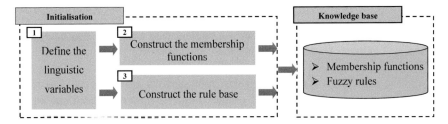

Fig. 3.6 Knowledge base construction of fuzzy logic approach

knowledge base, fuzzification, inference engine and defuzzification, as shown in Fig. 3.5.

The most important component in a FLS is the knowledge base. This component stores both the membership functions and the IF-THEN rules base provided by experts. Three steps, linguistic variables, membership functions and fuzzy rules, are prepared to establish a knowledge base (Abd et al. 2012a, b), as depicted in Fig. 3.6. The next sub-section will describe the previous three steps.

3.3.3 Linguistic Variables

A linguistic variable is defined as a variable whose values are not number, but words or sentences in natural language. In general, a linguistic variable consists of a set of words or phrases called linguistic terms, denoted by T. For example, if processing time is interpreted as a linguistic variable, to qualify the processing time terms such as "*Short*", "*Medium*" and "*Long*" processing time are used in a real industry context. These terms are called a fuzzy set of the processing time. Hence, a linguistic variable of processing time could be T [*Processing time*] = [*Short*, *Medium*, *Long*].

3.3.4 Membership Functions

A membership function (MF) embodies a fuzzy set Ã graphically. The values of the membership functions are between 0 and 1, denoted by μÃ (x) where x is an element of Ã; these values are called degree of membership. Figure 3.7 shows the most well-known of memberships functions shapes, namely triangular and trapezoidal (Mendel 1995).

A triangular fuzzy number can be defined by a triplet (a, b, c) and a trapezoidal fuzzy number is represented by four values (a, b, c, d) as shown in Eqs. 3.10 and 3.11.

$$\mu_A(x) = \begin{cases} 0, & x < a \\ \frac{x-a}{b-a}, & a \le x \le b \\ \frac{c-x}{c-b}, & b \le x \le c \\ 0, & x > c \end{cases} \qquad (3.10)$$

$$\mu_A(x) = \begin{cases} 0, & x < a \\ \frac{x-a}{b-a}, & a \le x \le b \\ 1, & b \le x \le c \\ \frac{d-x}{d-c}, & c < x \le d \\ 0, & x > d \end{cases} \qquad (3.11)$$

3.3.5 Fuzzy Rules

A fuzzy rule is structured to control the output variable. A fuzzy rule has two parts, the antecedent and the consequent: IF <antecedent> THEN <consequent>. For instance, IF x is A THEN y is B; where x and y are variables and A and B are linguistic variables determined by fuzzy sets.

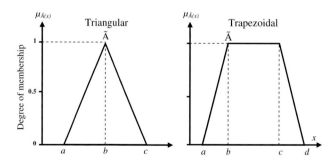

Fig. 3.7 Two examples of fuzzy numbers, triangular and trapezoidal

3.3.5.1 Simulation Module

Once the scheduling parameters, objective functions, constraints and decision variables are determined, the simulation module is defined and constructed. In this module, a computer simulation model of the RFAC is built, to evaluate the system performance under different scheduling strategies. In this research, simulation software called SIMPROCESS (Swegles 1997; CACI 2006) is used to build and simulate the assembling processes (Swegles 1997; CACI 2006). The process of simulation of RFAC is achieved through main four stages, using SIMPROCESS software. These stages are shown in Fig. 3.8.

The first stage is constructing a computer model of the RFAC. This stage is divided into three steps; define the model, construct a software model, and make a pilot run. Step one is based on the conceptual model that represents all the information related to the system, such as the components of the system and its layout, inputs required, assumptions, output generated. Step two is to construct the proposed model as a computer program; this can be done via encoding the mathematical and logical information of the system in a form that can be achieved by the computer software. After the model is defined and constructed, a pilot run is done in step three in order to be sure that the model is working as per the design requirements, and to detect any errors before beginning the simulation process. In SIMPROCESS, verification is an essential activity for checking the validity of the constructed model. Animation is another powerful activity for verifying the constructed model and visualizing the process in motion. The second stage as shown in Fig. 3.8 is running a model to generate the desired solutions. The model is run based on different numbers of experiments. In this research, the design experiments are determined by the output of the scheduling module, which represents the sequence of job flow to the system. The third stage is computing the performance measures. In this research, five performance measures are used. The last stage is evaluating alternative scenarios, in order to evaluate the RFAC performance under different scheduling strategies. The overall architecture of the proposed methodology is depicted in Fig. 3.9.

Fig. 3.8 Simulation process in SIMPROCESS

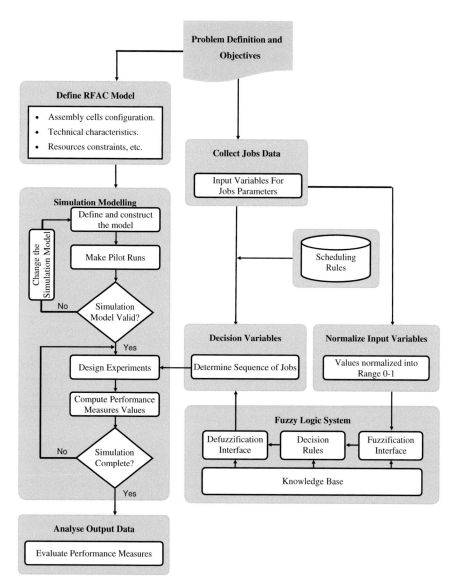

Fig. 3.9 Proposed strategy for scheduling problems in RFAC

3.4   Concluding Remarks

In this chapter, a new methodology for scheduling RFAC has been developed with the objective of minimizing makespan, robots idle time, total tardiness, maximum tardiness and number of tardy jobs. The developed methodology was divided into three main modules, namely pre-processing, scheduling and simulation.

- In the pre-processing module, the required components of modelling the scheduling problems in the RFAC were defined and described. These components were: parameters, objective functions, constraints and decision variables.
- In the scheduling module, the schedule for assembling multi products was generated via a new and sophisticated scheduling rule, namely fuzzy sequencing rule (FSR). FSR was constructed using a fuzzy-based mathematical model. This model used the membership functions to find the contribution of each product type to the output and then generate the sequence of products flow to the RFAC. The sequence generation was determined by normalization of each job variable such as processing time, batch size, due date and number of required stations.
- In the simulation module, a computer simulation model of the RFAC was built in SIMPROCESS, and then simulated under different scenarios, depending on the outcome of the scheduling module.

The methodology developed in this chapter will be proved in Chap. 4, using a realistic case study, and the simulation results will be compared with those of the common scheduling rules for evaluating the RFAC performance.

References

Abd, K., Abhary, K., & Marian, R. (2011). Scheduling and performance evaluation of robotic flexible assembly cells under different dispatching rules. *Advances in Mechanical Engineering, 1*(1), 31–40.

Abd, K., Abhary, K., & Marian, R. (2012a). Efficient scheduling rule for robotic flexible assembly cells based on fuzzy approach. In *Proceedings of the 45th CIRP Conference on Manufacturing Systems* (vol. 3, no. 3, pp. 483–488), Athens, Greece.

Abd, K., Abhary, K., & Marian, R. (2012b). Intelligent modeling of scheduling robotic flexible assembly cells using fuzzy logic. In *12th WSEAS International Conference on Robotics, Control and Manufacturing Technology* (pp. 202–207), Rovaniemi, Finland.

Abd, K., Abhary, K., & Marian, R. (2013a). Development of a fuzzy-simulation model of scheduling robotic flexible assembly cells. *Journal of Computer Science, 9*(12), 1761–1768.

Abd, K., Abhary, K., & Marian, R. (2013b). Intelligent model of scheduling RFACs—part I: Methodology and strategy. In *DAAAM international scientific book* (pp. 719–736). Vienna: DAAAM International Publishing.

Abd, K., Abhary, K., & Marian, R. (2013c). *A methodology for scheduling robotic flexible assembly cells using fuzzy logic and simulation* (pp. 449–454). London, UK: Proceedings of the World Congress on Engineering.

Baptiste, P., & Schieber, B. (2003). 'A note on scheduling tall/small multiprocessor tasks with unit processing time to minimize maximum tardiness. *Journal of Scheduling, 6*(4), 395–404.

Berrichi, A., & Yalaoui, F. (2013). Efficient bi-objective ant colony approach to minimize total tardiness and system unavailability for a parallel machine scheduling problem. *The International Journal of Advanced manufacturing Technology, 68*(9), 2295–2310.

Bilkay, O., Anlagan, O., & Kilic, S. E. (2004). Job shop scheduling using fuzzy logic. *International Journal of Advanced Manufacturing Technology, 23*(7), 606–619.

Buil, R., Piera, M. A., & Luh, P. B. (2010). Improvement of lagrangian relaxation convergence for production scheduling. *IEEE Transactions on Automation Science and Engineering, 9*(1), 137–147.

CACI. (2006). *User's manual: Simprocess*. La Jolla, CA: CACI Products Company.

Canbolat, Y. B., & Gundogar, E. (2004). Fuzzy priority rule for job shop scheduling. *Journal of Intelligent Manufacturing, 15*(4), 527–533.

Chan, T. S. F., & Chan, K. H. (2004). A comprehensive survey and future trend of simulation study on FMS scheduling. *Journal of Intelligent Manufacturing, 15*(1), 87–102.

Danping, L., & Lee, C. K. M. (2010). A review of the research methodology for the re-entrant scheduling problem. *International Journal of Production Research, 49*(8), 2221–2242.

Desal, N. K. (1997). *Scheduling algorithm for flexible manufacturing cells.* Master of Science, University of Manitoba.

Domingos, J. C., & Politano, P. R. (2003). On-line scheduling for flexible manufacturing systems based on fuzzy logic. *Proceedings of the IEEE International Conference on Systems, Man and Cybernetics, 5,* 4928–4933.

Jayamohan, M. S., & Rajendran, C. (2000). New dispatching rules for shop scheduling: A step forward. *International Journal of Production Research, 38*(3), 563–586.

Kumar, R. R., Singh, A. K., & Tiwari, M. K. (2004). A fuzzy based algorithm to solve the machine-loading problems of a FMS and its neuro fuzzy Petri net model. *International Journal of Advanced Manufacturing Technology, 23*(5), 318–341.

Mahdavi, I., Fekri Moghaddam Azar, A. H., & Bagherpour, M. (2009). Applying fuzzy rule based to flexible routing problem in a flexible manufacturing system. In *Proceedings of the IEEE International Conference on Industrial Engineering and Engineering Management* (pp. 2358–2364).

Marian, R. M., Kargas, A., Luong, L. H. S., & Abhary, K. (2003). A framework to planning robotic flexible assembly cells. In *32nd International Conference on Computers and Industrial Engineering* (pp. 607–615), Limerick, Ireland.

Mendel, J. M. (1995). Fuzzy logic systems for engineering: A tutorial. *IEEE, 83,* 345–377.

Ramasesh, R. (1990). Dynamic job shop scheduling: A survey of simulation research. *OMEGA. International Journal of Management Science, 18*(1), 43–57.

Restrepo, I. M., & Balakrishnan, S. (2008). Fuzzy-based methodology for multi-objective scheduling in a robot-centered flexible manufacturing cell. *Journal of Intelligent Manufacturing, 19*(4), 421–432.

Sridhar, S., Prabaharan, T., & Saravanan, M. (2010). Optimisation of sequencing and scheduling in hybrid flow shop environment using heuristic approach. *International Journal of Logistics Economics and Globalisation, 2*(4), 331–351.

Srinoi, P., Minyong, P., Shayan, E., & Ghotb, F. (2008). Routing and sequencing determination in flexible manufacturing system using a fuzzy logic approach. *Asian International Journal of Science and Technology in Production and Manufacturing, 1*(2), 127–138.

Srinoi, P., Shayan, E., & Ghotb, F. (2006). A fuzzy logic modelling of dynamic scheduling in FMS. *International Journal of Production Research, 44*(11), 2183–2203.

Subramaniam, V., Ramesh, T., Lee, G. K., Wong, Y. S., & Hong, G. S. (2000). Job shop scheduling with dynamic fuzzy selection of dispatching rules. *International Journal of Advanced Manufacturing Technology, 16*(10), 759–764.

Swegles, S. (1997). Business process modeling with SIMPROCESS. In *Winter Simulation Conference* (pp. 606–610). Piscataway, NJ.

Tavakkoli-Moghaddam, R., Moslehi, G., Vasei, M., & Azaron, A. (2005). Optimal scheduling for a single machine to minimize the sum of maximum earliness and tardiness considering idle insert. *Applied Mathematics and Computation, 167*(2), 1430–1450.

Vidyarthi, N. K., & Tiwari, M. K. (2001). Machine loading problem of FMS: A fuzzy-based heuristic approach. *International Journal of Production Research, 39*(5), 953–957.

Xing, L. N., Chen, Y. W., Wang, P., Zhao, Q. S., & Xiong, J. (2010). A knowledge-based ant colony optimization for flexible job shop scheduling problems. *Applied Soft Computing, 10*(3), 888–896.

Chapter 4
Case Study 1: Application of the Developed Methodology Using Fuzzy Logic and Simulation

4.1 Introduction

The research work in Chap. 3 was to develop a methodology for scheduling RFAC in multi-product assembly environments by combining a fuzzy-based mathematical model with simulation modelling. In this chapter, the proposed methodology will be examined using a realistic case study to prove the effectiveness of the methodology in generating the schedule for assembling multi-products. Hence, the objective of this chapter is to demonstrate the application of this methodology. In order to achieve the stated objective, the following three steps have to be performed:

- Design and implement the fuzzy-based mathematical model using MATLAB fuzzy logic toolbox.
- Apply the proposed methodology for solving the scheduling problems in RFAC using a hypothetical case study.
- Analyze and compare the simulation results of the developed methodology in relation to the common scheduling policy.

4.2 A Fuzzy Logic Model for Scheduling RFAC

In this section, a fuzzy-based mathematical model is developed to combine all input fuzzy variables in one scheduling rule. The input fuzzy variables include processing time, batch size, due date, and number of required stations; the output fuzzy variable represents job priority. The proposed fuzzy-based mathematical model can be built using three steps (described in Sect. 3.3.2): first, defining the linguistic variables; second, constructing membership functions; and third, constructing fuzzy rules. The next sub-section will explain in detail how these steps are performed.

© Springer International Publishing Switzerland 2016
K.K. Abd, *Intelligent Scheduling of Robotic Flexible Assembly Cells*,
Springer Theses, DOI 10.1007/978-3-319-26296-3_4

4.2.1 Defining the Linguistic Variables

The first step is to define the linguistic inputs/output variables. Each linguistic variable is divided into a set of linguistic terms. For instance, if processing time is interpreted as a linguistic variable, terms such as *short*, *medium* and *long* processing time are used in a real industry context to quantify the processing time. In this model, it is assumed as an example that processing time, due date and batch size have three linguistic variables, number of required stations has two linguistic variables, while the output variable, product priority, has seven linguistic variables, as shown in Table 4.1.

4.2.2 Constructing Membership Functions

In this model, the input/output variables are constructed from triangular and/or trapezoidal shapes which, as mentioned earlier, are the most well-known of membership functions. Processing time is constructed as a triangular shape. The processing time value is evaluated based on the time of all required tasks needed to complete the job. The categories in the evaluated value of processing time are low, medium and long, as shown in Fig. 4.1.

Batch size is also constructed as a triangular shape. The batch size value is measured and evaluated through the required quantity of each job. There are three

Table 4.1 Input and output variables with their fuzzy values

System variable	Linguistic variable	Linguistic value	Term set
Input	Processing time	Short	S
		Medium	M
		Long	L
	Batch size	Small	S
		Medium	M
		Large	L
	Due date	Short	S
		Medium	M
		Long	L
	Number of required stations	Low	L
		High	H
Output	Job priority	Very low	VL
		Low	L
		Below average	BA
		Average	A
		Above average	AA
		High	H
		Very high	HV

Fig. 4.1 Membership
function for normalized
"Processing Time" input

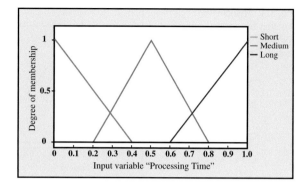

categories in the quantity of batch size, namely small, medium and large, as shown
in Fig. 4.2.

Due date is built from both triangular and trapezoidal shapes. The due date value
is measured based on the production deadline for each job. The categories in the
assessed value of due date are *small*, *medium* and *large*, as depicted in Fig. 4.3.

Fig. 4.2 Membership
function for normalized
"Batch Size" input

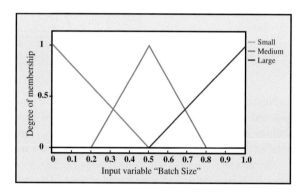

Fig. 4.3 Membership
function for normalized "Due
Date" input

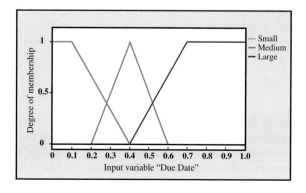

Fig. 4.4 Membership function for normalized "Number of Required Stations" input

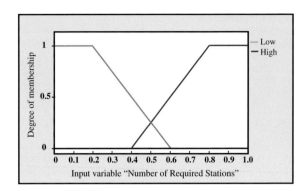

Fig. 4.5 Membership function for normalized "Job Priority"

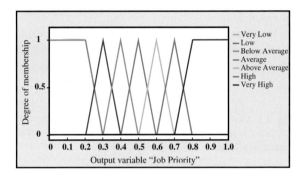

Number of required stations is constructed as a trapezoidal shape. The number of required stations value is measured and assessed based on the number of essential stations required for completing the job tasks. There are two categories in the number of required stations, namely *low* and *high* as shown in Fig. 4.4.

The product priority, which represents the fuzzy output of the suggested model, takes both triangular and trapezoidal shapes. The product priority is assessed and measured based on the priority status of products to be sequenced, from the highest product priority to the lowest order priority. The categories in the assessed value of product priority are *very low*, *low*, *below average*, *average*, *above average*, *high* and *very high*, as depicted in Fig. 4.5.

4.2.3 Constructing Fuzzy Rules

Fuzzy rules are structured to control the output variable. These rules can be provided by experts or may be extracted from numerical data. Since the variables of processing time, batch size and due date have three states each and the number of required stations has two states, the total number of fuzzy rules adds up to fifty four

Table 4.2 Fuzzy rules matrix

Number of station "High"		Processing time			Number of station "Small"		Processing time		
		Short	Medium	Long			Short	Medium	Long
Batch size "Small"									
Due date	Short	VH	VH	H	Due date	Short	AA	AA	A
	Medium	H	H	AA		Medium	A	A	BA
	Long	H	AA	A		Long	A	BA	BA
Batch size "Medium"									
Due date	Short	VH	H	AA	Due date	Short	A	A	BA
	Medium	AA	AA	A		Medium	BA	BA	L
	Long	AA	A	A		Long	BA	L	VL
Batch size "Large"									
Due date	Short	AA	AA	A	Due date	Short	BA	BA	L
	Medium	A	BA	BA		Medium	L	VL	VL
	Long	A	BA	BA		Long	L	VL	VL

($3 \times 3 \times 3 \times 2 = 54$), demonstrated in Table 4.2. The generic form of a fuzzy rule can be stated in the following form: IF (*Processing Time* is ■) and (*Due Date* is ■) and (*Batch Size* is ■) and (*Number of Required Stations* is ■) THEN (*Priority* is ■). The black boxes represent the linguistic variables for each of the fuzzy variables. The fuzzy rules derived are shown as in the example below.

1. IF (*Processing Time is S*) and (*Due Date is S*) and (*Batch Size is S*) and (*Number of Required Stations is H*) THEN (*Priority is VH*).
2. 2. IF (*Processing Time is M*) and (*Due Date is S*) and (*Batch Size is S*) and (*Number of Required Stations is H*) THEN (*Priority is VH*).
3. IF (*Processing Time is L*) and (*Due Date is S*) and (*Batch Size is S*) and (*Number of Required Stations is L*) THEN (*Priority is H*).

⋮

54. IF (*Processing Time is S*) and (*Due Date is S*) and (*Batch Size is S*) and (*Number of Required Stations is L*) THEN (*Priority is VL*).

4.3 Implementation of Fuzzy Approach for Scheduling of RFAC

In this section, the proposed scheduling rule (FSR) is implemented using MATLAB fuzzy logic toolbox (Abd et al. 2012a, b). The fuzzy logic toolbox consists of five graphical user interface tools (GUIs) for building, editing and observing any fuzzy inference system (Sivanandam et al. 2007; Mathworks 2009). These tools are: the

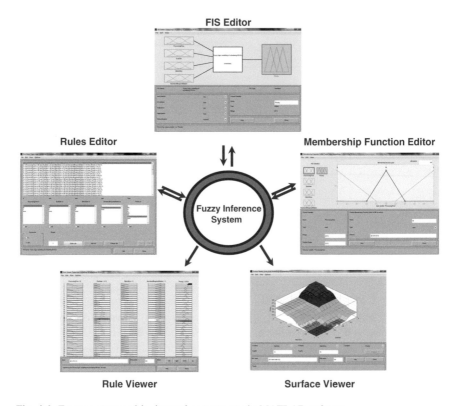

Fig. 4.6 Fuzzy system and its integral components in MATLAB software

fuzzy inference system (FIS) editor, the membership function editor, the rule editor, the rule viewer, and the surface viewer, as shown in Fig. 4.6. The GUIs are dynamically connected, and the altering of any GUI can affect the other GUIs.

In the fuzzy logic toolbox, the fuzzy inference system (FIS) editor handles the information related to the variables of inputs and output, such as the variables' names and their numbers. In this research, four fuzzy input variables, processing time, batch size, due date and number of required stations, are defined. The job priority is the fuzzy output variable, representing the priority status of the product to be selected for the next assembly operation in the RFAC. The membership function editor is used to construct the shapes of all the input/output variables. The membership function shape of each fuzzy variable was described in Sect. 4.2.2.

The rule editor is for editing the list of fuzzy rules that are used to control the output variable. The fuzzy rule is constructed based on the number of linguistic variables for inputs/output (see Sect. 4.2.3 for definition).

The surface viewer allows the user to visualize the relation between input fuzzy variables and the output of a fuzzy system in a three-dimensional graph; the X-axis and Y-axis in the 3D graph represent any two selected input variables, and the Z-axis represents the output of a fuzzy system. In this research, four input variables

Fig. 4.7 Output surfaces of
the FIS for processing time
and due date

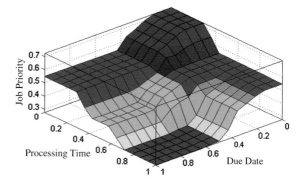

Fig. 4.8 Output surfaces of
the FIS for processing time
and batch size

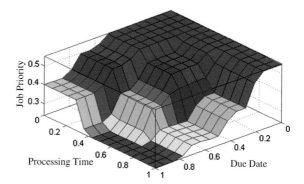

are used, so the number of generated 3D graphs is six, as shown in Figs. 4.7, 4.8,
4.9, 4.10, 4.11 and 4.12.

Figure 4.7 illustrates the priority resulting from the interaction of processing time
and due date. From this figure it can be seen that the short processing time and
small due date values give a high score of product priority. Moreover, it can be seen
that the processing time has a slightly higher influence than the due date on the
product priority.

Fig. 4.9 Output surfaces of
the FIS for processing time
and number of required
stations

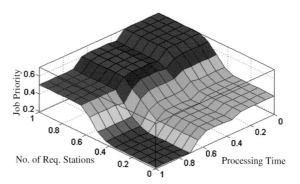

Fig. 4.10 Output surfaces of the FIS for batch size and number of required stations

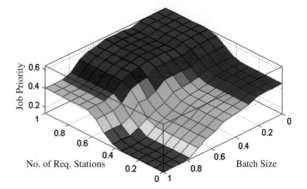

Fig. 4.11 Output surfaces of the FIS for batch size and due date

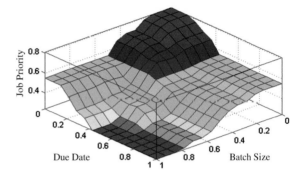

Figure 4.8 depicts the priority response surface for processing time and batch size. It can be seen that the batch size has a higher influence than the processing time on the product priority. Again, a short processing time and small batch size signify a high priority score.

Figure 4.9 represents the job priority from the perspective of processing time and number of required stations. Again a short processing time, regardless of the number of required stations, gives a high job priority score. Also, it can be seen that the processing time has a smaller impact on job priority evaluation as compared to the number of required stations.

Figure 4.10 depicts the priority resulting from the interaction of batch size and number of required stations. From this figure it can be concluded that the small batch size and high number of required stations give a high job priority score.

Figure 4.11 illustrates the job priority from the perspective of batch size and due date. As shown in this figure, the batch size has a greater influence on the job priority than the due date. It is also shown that the job priority decreases with decreasing batch size and due date.

Figure 4.12 depicts the priority response surface with due date and number of required stations, from which it can be inferred that the small due date and high number of required stations gives a high job priority score.

Fig. 4.12 Output surfaces of the FIS for due date and number of required stations

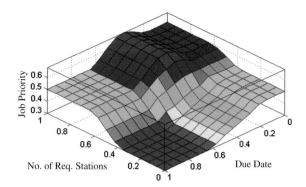

4.4 Example Application of Scheduling RFAC

4.4.1 RFAC Description

The RFAC studied in this chapter consists of three main components: (1) Robots (R_1 and R_2) which fetch the required parts and place them at assembly stations (S_1, S_2 and S_3) where the parts are assembled. (2) Part feeder (PF) which supplies parts to the cell. (3) Input and output conveyors (IC and OC) which supply the base parts and carry out the final products, as depicted in Fig. 4.13.

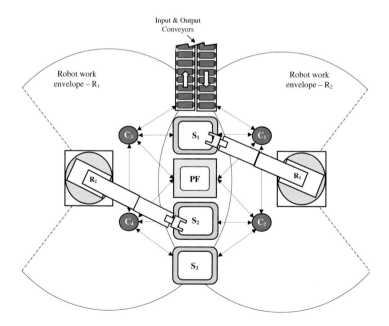

Fig. 4.13 A robotic flexible assembly cell

The following assumptions and limitations are considered in the simulation model to provide a reliable solution to practical cases:

- The optimum assembly sequence of each product is given in advance.
- Each product uses some or all of the cell resources.
- Each robot can perform only one task at a time.
- Each robot has multi-purpose end effectors.
- The processing time of each task is deterministic and is known in advance.

In this system, four control points $\{C_1, C_2,..., C_4\}$ are set to simplify path planning and avoid collisions between robots in the shared area. Table 4.3 shows the robot paths and their required time to move between two positions in the cell.

The RFAC described above is assumed to assemble n product types $(P_1, P_2,..., P_n)$. Each product is considered as an independent job. In this case study, six products are taken as an example. Table 4.4 shows the details of required stations along with the assembly operations time for each product type. This table also includes parts pick up and release times for the robots assembling the products.

In order to simulate RFAC, three customer orders are assumed and labelled as order #1, #2 and #3, as shown in Table 4.5. Orders #1 and #3 consist of six types of cell phone, and order #2 is composed of only five types of cell phone. Batch size and due date for each product type are also given in Table 4.5.

Table 4.3 Transportation time for robots between cell resources

Path description	Position	Travel time
Robot move from resource to control point	S_1, PF $\rightarrow C_1$, C_3 S_2, S_3, PF $\rightarrow C_2$, C_4	0.5
Robot move from control point to resource	C_1, $C_3 \rightarrow S_1$, PF C_2, $C_4 \rightarrow S_2$, S_3, PF	1
Robot move between control point and conveyor	C_1, $C_3 \leftrightarrow$ IC C_1, $C_3 \leftrightarrow$ OC	1.5
Robot move between two control points	$C_1 \leftrightarrow C_2$ $C_3 \leftrightarrow C_4$	0.5

Table 4.4 Assembly operations requirements

Description	Assembly station	Time of assembly operations (s)					
		P_1	P_2	P_3	P_4	P_5	P_6
Insert lens on front cover	S_1	4	3	3	4	3	4
Insert keypad on front cover	S_1	5	4	5	6	4	6
Attach PC board to front cover	S_2	6	8	10	9	8	9
Insert antenna on back cover	S_3	9	0	0	9	0	0
Attach back cover to front cover	S_2	7	11	10	11	7	10
Robot gripper pickup and release time (s)		6	4	4	6	4	4

Table 4.5 Orders for product types with different production volume

Product type	Order #1		Order #2		Order #3	
	Batch size	Due date	Batch size	Due date	Batch size	Due date
P_1	3	450	2	1200	4	1500
P_2	6	650	6	1300	5	1900
P_3	5	800	5	1400	3	1650
P_4	3	600	3	1000	3	1700
P_5	5	400	4	1100	3	1850
P_6	6	500	–	–	4	2000
Prod. volume	28		20		22	

Table 4.6 shows the values of the membership functions for input data. These values can be determined by evaluating the normalization of each job variable such as processing time, batch size, due date and number of required stations. The normalization can be done, using Eqs. 3.6–3.9. Following is an example of a calculation for P_1 in Orders #1:

$$\mu_T^1 = \frac{[180 - 180]}{[300 - 180]} = 0, \quad \mu_D^1 = \frac{[450 - 400]}{[800 - 400]} = 0.13, \quad \mu_N^1 = \frac{[3 - 3]}{[6 - 3]} = 0,$$
$$\mu_S^1 = \frac{[3 - 2]}{[3 - 2]} = 1$$

Table 4.6 Membership functions for input data and priority values

O_h	P_i	P_i	N_i	D_i (s)	S_i	μ_T^i	μ_D^i	μ_N^i	μ_S^i
Order #1	P_1	180	3	450	3	0	0.13	0	1
	P_2	282	6	650	2	0.85	0.63	1	0
	P_3	245	5	800	2	0.54	1	0.67	0
	P_4	204	3	600	3	0.20	0.50	0	1
	P_5	215	5	400	2	0.29	0	0.67	0
	P_6	300	6	500	2	1	0.25	1	0
Order #2	P_1	120	2	1200	3	0	0.50	0	1
	P_2	282	6	1300	2	1	0.75	1	0
	P_3	245	5	1400	2	0.77	1	0.75	0
	P_4	204	3	1000	3	0.52	0	0.25	1
	P_5	172	4	1100	2	0.32	0.25	0.50	0
Order #3	P_1	240	4	1500	3	1	0	0.50	1
	P_2	235	5	1900	2	0.95	0.80	1	0
	P_3	147	3	1650	2	0.16	0.30	0	0
	P_4	204	3	1700	3	0.68	0.40	0	1
	P_5	129	3	1850	2	0	0.70	0	0
	P_6	200	4	2000	2	0.64	1	0.50	0

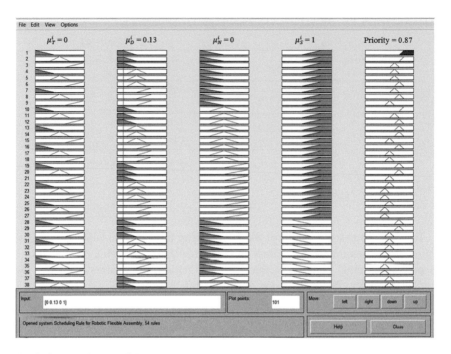

Fig. 4.14 Rule viewers of the FIS

The rule viewer, which displays a graphical representation of the values of the input variables and the output of a fuzzy system through all the fuzzy rules, is shown as an example in Fig. 4.14. The output (job priority) in this figure can be interpreted easily as follows: IF the μ_T^i is (0), the μ_D^i is (0.13), the μ_N^i is (0), and the μ_S^i is (1) THEN *priority* will be (0.87). Table 4.7 shows the final results of the job prioritizing.

4.4.2   Simulation of Experimental Design

In this section, the experimental design is set. Each experiment is performed with a different scheduling rule. Seven experiments are implemented. Experiments numbered 1–6 are run with existing scheduling rules; Experiment 7 is run using the developed rule. The selected rules generate different sequences of product flow to the system, as shown in Table 4.8.

Table 4.7 Membership functions for inputs data and priority values

O_h	P_i	μ_T^i	μ_D^i	μ_N^i	μ_S^i	Priority	Sequence
Order #1	P_1	0	0.13	0	1	0.87	1
	P_2	0.85	0.63	1	0	0.13	6
	P_3	0.54	1	0.67	0	0.20	5
	P_4	0.20	0.50	0	1	0.60	2
	P_5	0.29	0	0.67	0	0.45	3
	P_6	1	0.25	1	0	0.22	4
Order #2	P_1	0	0.50	0	1	0.60	2
	P_2	1	0.75	1	0	0.12	5
	P_3	0.77	1	0.75	0	0.14	4
	P_4	0.52	0	0.25	1	0.82	1
	P_5	0.32	0.25	0.50	0	0.45	3
Order #3	P_1	1	0	0.50	1	0.70	1
	P_2	0.95	0.80	1	0	0.13	6
	P_3	0.16	0.30	0	0	0.59	3
	P_4	0.68	0.40	0	1	0.67	2
	P_5	0	0.70	0	0	0.50	4
	P_6	0.64	1	0.50	0	0.24	5

Table 4.8 Designing a simulation experiment

No.	Experiment	Order no.	Sequence
1	Short processing time (*SPT*)	#1	1–4–5–3–2–6
		#2	1–3–5–4–2
		#3	5–3–6–4–2–1
2	Long processing time (*LPT*)	#1	6–2–3–5–4–1
		#2	2–4–5–3–1
		#3	1–2–4–6–3–5
3	Random (*RAN*)	#1	3–2–4–6–1–5
		#2	3–2–1–5–4
		#3	3–1–2–5–6–4
4	Earlier due date (*EDD*)	#1	5–1–6–4–2–3
		#2	4–5–1–2–3
		#3	1–3–4–5–2–6
5	Critical ratio (*CR*)	#1	6–5–1–2–4–3
		#2	4–5–2–1–3
		#3	1–4–3–2–5–6
6	Minimise slack time (*MST*)	#1	5–6–1–4–2–3
		#2	4–5–2–1–3
		#3	1–3–4–2–5–6
7	Fuzzy sequencing rule (*FSR*)	#1	1–4–5–6–3–2
		#2	4–1–5–3–2
		#3	1–4–3–5–6–2

4.5   Simulation Results and Discussion

In this section, the results of the simulation study are discussed. The discussion will focus on analyzing the results and comparing the RFAC performance based on the proposed rule (FSR) and existing scheduling rules. Five common performance measures, namely makespan, percentage of idle time, total tardiness, maximum tardiness and percentage of tardy jobs, are used to determine the performance of the RFAC (Abd et al. 2013a, b). As mentioned earlier, three customer orders, as shown in Table 4.5, are assumed in order to simulate RFAC. The simulation results of the overall performance measures of the three different orders and the average results are presented in Tables 4.9, 4.10, 4.11, 4.12 and 4.13. The comparisons of all scheduling rules with respect to the five performance measures are shown in Figs. 4.15, 4.16, 4.17, 4.18 and 4.19.

Table 4.9 Simulation results of makespan for different orders

Order number	Scheduling rules						
	SPT	LPT	RAND	EDD	CR	MST	FSR
Order #1	763.5	760.5	790.5	789.5	786.5	765.5	717.5
Order #2	1287	1326	1353	1358	1356	1335	1239
Order #3	1920	1990.5	1997.5	2022.5	2020	2005.5	1839
Av. makespan	1323.5	1359.0	1380.3	1390.0	1387.5	1368.7	1265.2

Table 4.10 Simulation results of percentage of robots idle time for different orders

Order number	Scheduling rules						
	SPT (%)	LPT (%)	RAND (%)	EDD (%)	CR (%)	MST (%)	FSR (%)
Order #1	7.0	6.0	10.0	10.0	9.0	7.0	1.0
Order #2	5.0	8.0	9.0	10.0	10.0	8.0	1.0
Order #3	6.0	9.0	10.0	11.0	11.0	10.0	2.0
Av. idle time	6.0	7.7	9.7	10.3	10.0	8.3	1.3

Table 4.11 Simulation results of total tardiness for different orders

Order number	Scheduling rules						
	SPT	LPT	RAND	EDD	CR	MST	FSR
Order #1	263.5	543	710	0	0	0	67.5
Order #2	181	288.5	464	0	69.5	51.5	0
Order #3	420	440	381.5	58.5	141.5	63.5	0
Av. total tardiness	288.2	423.8	518.5	19.5	70.3	38.3	22.5

Table 4.12 Simulation results of maximum tardiness for different orders

Order number	Scheduling rules						
	SPT	LPT	RAND	EDD	CR	MST	FSR
Order #1	263.5	310.5	390.5	0	52	0	67.5
Order #2	181	126	353	0	69.5	51.5	0
Order #3	420	255.5	297.5	24.5	63	52.5	0
Av. max. tardiness	288.2	230.7	347.0	8.2	61.5	34.7	22.5

Table 4.13 Simulation results of percentage of tardy jobs for different orders

Order number	Scheduling rules						
	SPT (%)	LPT (%)	RAND (%)	EDD (%)	CR (%)	MST (%)	FSR (%)
Order #1	17.0	50.0	50.0	0.0	17.0	0.0	17.0
Order #2	20.0	60.0	40.0	0.0	20.0	20.0	0.0
Order #3	17.0	50.0	33.0	33.0	67.0	50.0	0.0
Av. % of tardy jobs	18.0	53.0	41.0	11.0	35.0	23.0%	6.0

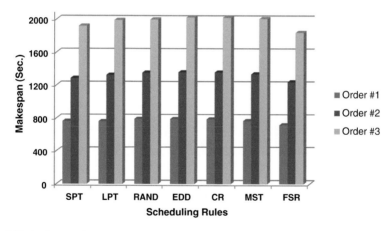

Fig. 4.15 Makespan for all orders

One of the important measures of manufacturing system performance is make-span which represents the maximum completion time for the entire set of jobs. Shorter makespan results in due dates of customer orders being met, as well as a decrease in the direct production cost. Figure 4.15 shows the makespan results of scheduling rules for different customer orders. It can be seen that the developed rule (FSR) obtains the best results for minimizing the makespan, compared with the other scheduling rules. SPT and LPT rank second and third respectively. CR and EDD are the worst in minimizing the makespan objective, for the reason that CR

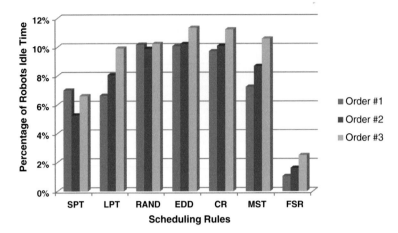

Fig. 4.16 Percentage of robots idle time for all orders

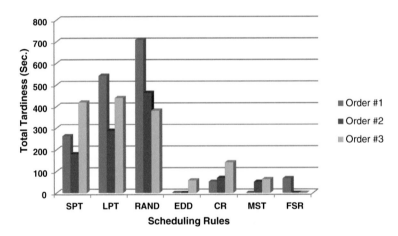

Fig. 4.17 Total tardiness for all orders

and EDD concentrate only on due dates of jobs and ignore the other variables such as processing time and batch size.

Robots idle time is an important time-based measure for scheduling evaluation. Since robots are costly investments, it is vital to use them efficiently by reducing the idle time. This criterion enables a clear evaluation as to whether the robots are used in an efficient way. Figure 4.16 shows the percentage of idle time of scheduling rules on the three orders. In this figure, FSR emerges as the best rule among all seven scheduling rules, followed by SPT and LPT. SPT and LPT give good results for this measure. EDD appears to be the worst rule for minimizing the robots' idle time. The reason for the poor performance of this rule is that the EDD rule

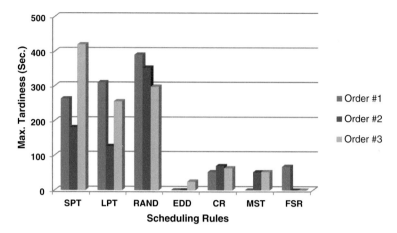

Fig. 4.18 Maximum tardiness for all orders

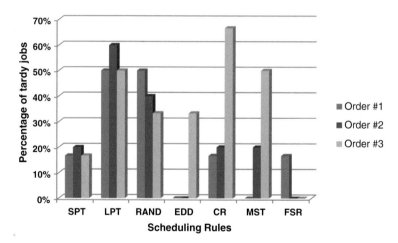

Fig. 4.19 Percentages of tardy jobs for all orders

concentrates only on the due date for the complete set of jobs and ignores the variable of processing time.

Total tardiness is another performance measure typically used in scheduling evaluation. This criterion represents the summation of jobs that fail to meet the due date. A higher total tardiness may result in loss of customers and competitiveness, as penalty for the late completion. The overall total tardiness of the scheduling rules on the three different orders is depicted in Fig. 4.17, where EDD appears to be the best rule among all seven scheduling rules. The second rank goes to the FSR. The difference between the results of EDD and FSR is insignificant. SPT, LPT and

RAND are the worst in minimizing the total tardiness criteria. This is because the due date variable is ignored by these rules.

Maximum tardiness is another criterion related to the jobs' due dates. Maximum tardiness represents the largest difference between the final fulfilment date for jobs and the requested delivery date. Figure 4.18 shows the maximum tardiness of the scheduling rules on all three orders. The EDD rule emerges as the best for minimizing the maximum tardiness of the scheduling rules. FSR obtains a slightly lower performance compared to EDD. MST and CR rank third and fourth respectively. LPT, SPT and RAND are observed to be the worst performing rules for this measure, again for the reason that the due date is ignored by LPT, SPT and RAND; also LPT and SPT focus only on the processing time of jobs.

Like total tardiness and maximum tardiness, the percentage of tardy jobs is criterion-related to the jobs' due dates. This criterion is defined as the ratio of the number of tardy jobs to the number of jobs. The overall percentages of tardy jobs for scheduling rules on different orders are presented in Fig. 4.19. According to which FSR obtains the best results for minimizing the percentage of tardy jobs. EDD ranks second, and still obtains good results among all the other selected rules from the previous literature. LPT and RAND perform poorly in minimizing the percentages of tardy jobs in almost all orders. The poor performance of these rules is expected because they do not deal with the due date or batch size of jobs.

In terms of time based measures, the simulation results show that the developed rule (FSR) outperforms all the other selected rules from literature. SPT and LPT obtain acceptable performance. EDD is observed to be the worst performing rule for time based measures. With respect to due date based measures, it can be seen that the EDD rule appears to be the best in minimizing these measures; on the other hand it performs poorly in minimizing time based measures. Also, FSR proves very effective in minimizing the due date based measures.

From the above discussion, it can be concluded that FSR is generally better than all other common scheduling rules. This is because the developed rule is constructed by combining all input variables such as processing time, due date and batch size.

4.6 Concluding Remarks

This chapter is dedicated to the application of the methodology developed in Chap. 3. A hypothetical case study was used to demonstrate the effectiveness of this methodology for scheduling RFAC in a multi-product assembly environment. The scenario presented is realistic and of an average level of difficulty. The running time of this scenario is approximately 33 min. The results revealed that:

- The performance of the proposed fuzzy-based mathematical model was more efficient compared to the heuristic scheduling rules (SPT, LPT, RAND, EDD,

CR and MST), based on all five objectives: minimizing makespan, robots idle time, total tardiness, maximum tardiness and number of tardy jobs.
- The fuzzy-based mathematical model tackles the deficiencies in the heuristic scheduling rules by considering all the important input variables in the scheduling problems: processing time, batch size, due date and number of required stations.
- The heuristic scheduling rules were not suitable for finding an acceptable schedule regarding multi-objective criteria. For example EDD was efficient at due date based objectives; nevertheless it was worst with time-based objectives.

Even though the developed methodology was devoted to scheduling RFAC in a multi-product assembly environment, it concentrated only on the static scheduling problem. In fact, the real context of advanced manufacturing systems is often dynamic during the scheduling processes. Accordingly, the following chapters attempt to expand the developed methodology by considering the dynamic scheduling problems of RFAC.

References

Abd, K., Abhary, K., & Marian, R. (2012a). Efficient scheduling rule for robotic flexible assembly cells based on fuzzy approach. In *Proceedings of the 45th CIRP Conference on Manufacturing Systems* (Vol. 3, No. 3, pp. 483–488). Athens, Greece.

Abd, K, Abhary, K., & Marian, R. (2012b). Intelligent modeling of scheduling robotic flexible assembly cells using fuzzy logic. In *12th WSEAS International Conference on Robotics, Control and Manufacturing Technology* (pp. 202–207). Rovaniemi, Finland.

Abd, K., Abhary, K., & Marian, R. (2013a). Application of a fuzzy-simulation model of scheduling robotic flexible assembly cells. *Journal of Computer Science, 9*(12), 1769–1777.

Abd, K., Abhary, K., & Marian, R. (2013b). *Intelligent model of scheduling RFACs—part II: Application* (pp. 737–750). Vienna: DAAAM International Scientific Book, DAAAM International Publishing.

Mathworks. (2009). Fuzzy logic toolbox user's guide. http://www.mathworks. com/access/helpdesk/help/pdf_doc/fuzzy/fuzzy.pdf

Sivanandam, S. N., Sumathi, S., & Deepa, S. N. (2007). *Introduction to fuzzy logic using MATLAB*. New York: Springer.

Chapter 5
Simulation Modelling and Analysis of Dynamic Scheduling in RFAC

5.1 Introduction

In real industrial situations, the manufacturing systems may face unexpected events such as order cancellation, arrival of urgent orders, due date changing and unavailability of tools or materials. These events may lead to deviations from the schedule plan. Hence, the influential events must be taken into consideration when dealing with real-world scheduling problems (Rajabinasab and Mansour 2011; Ouelhadj and Petrovic 2009). The scheduling research work that takes into account such unexpected events is termed *dynamic scheduling* (Zhang et al. 2013; Rajabinasab and Mansour 2011; Ouelhadj and Petrovic 2009). The dynamic scheduling of manufacturing systems has received significant attention over the past decade due to its potential for employing these systems in more efficient ways (Xiang and Lee 2008; Ouelhadj and Petrovic 2009).

In this chapter, the dynamic scheduling of RFAC will be examined and discussed. The primary objective is to develop an intelligent approach for dynamic scheduling problems. This approach will address the shortcomings of the static scheduling problem discussed in Chap. 3. In order to handle the complexity of dynamic scheduling in RFAC, this chapter intends to:

- Review the literature on the dynamic events that should be considered in order to solve scheduling problems in manufacturing systems.
- Develop a new approach for dynamic scheduling problems in RFAC. This approach is based on computer simulation modelling and Taguchi method.
- Demonstrate the applicability of the developed approach via a realistic case study.

© Springer International Publishing Switzerland 2016
K.K. Abd, *Intelligent Scheduling of Robotic Flexible Assembly Cells*,
Springer Theses, DOI 10.1007/978-3-319-26296-3_5

5.2   Review of Literature on Dynamic Events

In real life, manufacturing systems are mostly operating in dynamic environments. These systems are often subject to factors that may cause deviations from the generated schedules, and the schedule plan may become impractical to implement when it is released to the system (Ouelhadj and Petrovic 2009). Hence, the influential factors must be taken into consideration when dealing with dynamic problems in any given production system (Jain and Elmaraghy 1997; Gholami et al. 2009). According to Zhang et al. (2013), Caprihan et al. (2013), Ali and Wadhwa (2010) and Chan et al. (2008), *sequencing rules* and *dispatching rules* have been considered as important factors to execute dynamic scheduling strategies in flexible manufacturing systems.

The other influential factors are categorized by some authors into two types (Vieira et al. 2003; Cowling and Johansson 2002; Stoop and Weirs 1996): the first is *resource-related* such as resource breakdown, tool failure, loading limits, shortage of material; the second is *job-related*, such as due date changing, early or late arrival time of jobs, changing of processing time, urgency of jobs, job cancellation and so on. Tang et al. (2005), Vinod and Sridharan (2008), Kianfar et al. (2009), Lu and Liu (2011), Rajabinasab and Mansour (2011), Nie et al. (2013) indicated that the dynamic jobs' *arrival time* and *due date* changing are the most significant factors that affect job shop scheduling. Hence, the current research will take into consideration those unexpected events that can be categorized as *job-related*, such as due date changing and early or late arrival time of jobs.

Arrival time: The arrival time is defined as the degree of workload in the system. A low workload means a long arrival time for jobs. Conversely, a high workload means a short arrival time. In recent studies of job shop scheduling, the arrival times have been assumed to follow an exponential distribution (Nie 2012; Lu and Liu 2011). The mean inter-arrival time of jobs is calculated using the following formula:

$$\propto \; = \frac{T_p \times N_g}{U \times M} \tag{5.1}$$

where $\propto$ is mean inter-arrival time for jobs, T_p is the mean processing time per job, N_g is mean number of operations per job, U is shop utilisation; and M is number of machines in the system.

Due date: The due date assignment problem has been reported by a number of researchers. A comprehensive review of the due date assignment procedure can be found in (Cheng and Gupta 1989; Kaplan and Unal 1993). In these studies, some of the common methods for due date assignment are described as follows:

$$\text{Constant (CON):} \quad D_i = A_i + K \tag{5.2}$$

$$\text{Random (RAN):} \quad D_i = A_i + e_i \tag{5.3}$$

Table 5.1 The factors reported in recent studies for the job and flow shop scheduling problems

Author	Year	Sequencing rules	Dispatching rules	Job arrivals	Due date changing	Other
Caprihan et al.	2013	✓	✓	×	✓	Information delay ratio and routing flexibility
Zhang et al.	2013	✓	✓	✓	✓	The quantity of new jobs
Nie et al.	2013	✓	×	✓	✓	Routing flexibility
Qiu and Lau	2013	×	✓	✓	✓	Machine efficiency and machine failure
Nie	2012	✓	✓	✓	✓	Routing flexibility
Rajabinasab and Mansour	2011	×	✓	✓	✓	Machine breakdown
Ali and Wadhwa	2010	✓	✓	×	×	Routing flexibility and number of pallets
Zhou et al.	2009	×	✓	✓	✓	×
Kianfar et al.	2009	×	✓	✓	✓	Number of stages in the shop
Chan et al.	2008	✓	✓	×	×	Review-period size and routing flexibility
Viond and Sridharan	2008	×	✓	✓	✓	Setup time ratio
Chan et al.	2007	✓	✓	×	×	Routing flexibility and physical and operating parameters
Chan et al.	2006	✓	✓	×	×	Penalty level varied and routing flexibility
Thiagarajan and Rajendran	2005	×	✓	✓	✓	×
Caprihan and Wadhwa	2005	✓	✓	×	×	Number of pallets, information delay and routing flexibility
Tang et al.	2005	×	✓	✓	✓	×
Dominic et al.	2004	×	✓	✓	✓	×

$$\text{Total Work Content (TWK):} \quad D_i = A_i + K \times TPT_i \tag{5.4}$$

$$\text{Slack (SLK):} \quad D_i = A_i + K + TPT_i \tag{5.5}$$

$$\text{Number of operations (NOP):} \quad D_i = A_i + K \times N_i \tag{5.6}$$

where D_i and A_i are the due date and arrival time of job i respectively, K is due date tightness factor, e_i is a random parameter, TPT_i is the total processing time of job i and N_i is the number of operations that job i will undergo.

From the above reviewed literature it can be seen that the *sequencing rules*, *dispatching rules*, *arrival time* and *due date* are the most important scheduling factors that must be considered in order to reflect scheduling problems of RFAC more realistically. Table 5.1 presents the factors selected by researchers for the scheduling problems.

5.3   A Framework for Developing an Intelligent Approach to Dynamic Scheduling Problems

In general, the scheduling process in advanced manufacturing systems is becoming increasingly complex and dynamic, and an intelligent approach is required to assist decision makers. As indicated earlier, the simulation-based approaches are promising approaches for the addressing of dynamic scheduling problems. Additionally, the Taguchi method is an efficient tool for setting the least number of possible experiments to solve scheduling problems and for analyzing the simulation results. Therefore, the objective of this part of the research is to present the simulation-based approach combined with Taguchi optimization method in a new application for dynamic scheduling problems in RFAC which takes into consideration the important scheduling factors influencing the job shop.

In this section, a framework is developed for scheduling problems of RFAC in dynamic environments (Abd et al. 2014a, b). The proposed approach aims to: (1) provide an efficient scheduling plan; (2) demonstrate the feasibility of combining the computer simulation model with Taguchi experimental design method to provide an effective approach for dynamic scheduling problems; (3) assist decision-makers to understand how the scheduling factors affect the RFAC performance in a dynamic environment, and (4) determine the optimal combination of selected scheduling factors to obtain the best results. The proposed approach comprises four phases: *preparation*, *application of Taguchi method*, *simulation modelling* and *statistical analysis*. These phases are described in the following sub-sections.

5.3.1   Preparation

In this phase, two steps are defined: identifying the scheduling problems and objectives, then determining scheduling factors and the number of levels for each factor. As stated in Chap. 3, the scheduling of the RFAC requires finding a way to determine how to use the cell resources in an optimal manner to assemble multi-products. RFAC can assemble a number of related products (a family of products) grouped according to group technology (GT) rules when the resources of RFAC deal with similar parts that have the same geometrical and physical characteristics.

Let us consider that customer orders $\{O = 1, 2, \ldots, m\}$ are processed in RFAC. Each order consists of a set of products $\{P = 1, 2, \ldots, n\}$. These products go through specific assembly operations $\{op = op_{1n}, op_{2n}, \ldots, op_{in}\}$ that have to be implemented by a set of robots $\{R = 1, 2, \ldots, t\}$. The robots have multi-purpose end effectors. Each robot can perform only one job at a time and each job can be processed by only one robot.

In this part of the research, three independent objectives, namely the minimizing of makespan (C_{max}), total tardiness (TD) and number of tardy jobs (N_T), are considered. These objectives are described using the following equations:

$$C_{max} = \max_{1 \leq i \leq p} \left(C_{PO}\right) \tag{5.7}$$

$$TD = \sum_{O=1}^{m} \sum_{P=1}^{n} \left(\max_{1 \leq p \leq n} [C_{PO} - D_{PO}, 0] \right) \tag{5.8}$$

$$N_T = \sum_{O=1}^{m} \sum_{P=1}^{n} (U_P), \quad U_P = \begin{cases} 1, & \text{if } C_{PO} > D_{PO} \\ 0, & \text{otherwise} \end{cases} \tag{5.9}$$

where C_{PO} and D_{PO} denote the completion time and due date of product P respectively in order O, and U_P is the indicator for whether product P is tardy or not.

The scheduling of any flexible manufacturing system contains a number of decision points that affect the system performance. These decision points can be identified via scheduling stages. At each decision point, different rules can be utilized for decision making. For example, the decision making for job selection usually includes the smallest number of jobs in the system's queue. Figure 5.1 shows two types of decision points, sequencing rules and dispatching rules, in the RFAC.

In this study, to reflect more of reality in scheduling problems, the decision points, namely sequencing rules and dispatching rules, are both considered when scheduling RFAC in a multi-product assembly environment.

Fig. 5.1 The decision points
in the RFAC

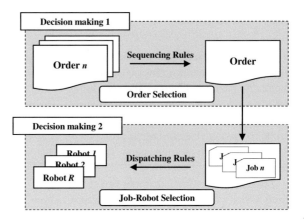

5.3.2   Application of Taguchi Method

The potential benefit of Taguchi method is its ability to solve complex problems with a drastically reduced number of experiments to be performed, accordingly reducing the cost of conducting the experiments (Ross 1988; Bendell et al. 1989). Taguchi designed special orthogonal arrays based on the number of factors and their levels. These arrays determine the number of experiments required. Taguchi suggested a robust design criterion called signal-to-noise (S/N) ratio. The S/N ratio characteristic is classified by Taguchi into three types: the smaller the better, the larger the better and the nominal the best (Lee 2000). There are two main motives for choosing Taguchi method. First, this method has been extensively used as an optimization method to meet actual design and manufacturing problems (Taguchi et al. 1989; Park 1996). Second, Taguchi method has been successfully used to conduct the experiments, analyze the simulation results and determine the significant factors that improve system performance (Caprihan et al. 2013; Moradi et al. 2010; Ali and Wadhwa 2010; Caprihan and Wadhwa 2005). In this phase, three steps are performed: firstly, selection of the appropriate Taguchi orthogonal array and assignment of scheduling factors to the orthogonal array; secondly, conducting of experiments based on the arrangement of the orthogonal array and measurement of the objective function values; thirdly, substitution of the objective function values into an S/N ratio value.

5.3.3   Simulation Modelling

In this section, simulation modelling is used to study the performance of the proposed scheduling policy. Simulation modelling of RFAC is developed and implemented using the software SIMPROCESS (see Appendix B). The simulation

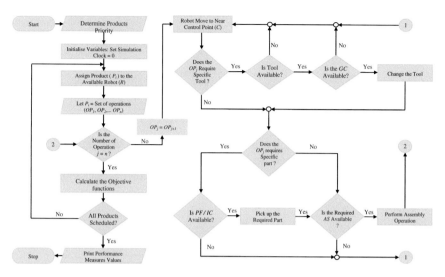

Fig. 5.2 Flow chart of the proposed algorithm

modelling is made up of the following two main components: *Activities*: the required tasks that are carried out by utilizing robots to assemble the products. These activities are: move, tool-change, pick-up, assembly. *Resources*: the physical elements of RFAC: robots (R), assembly stations (AS), parts feeder (PF), gripper changing station (GC), input/output conveyors (IC and OC). The proposed algorithm for scheduling products using the resources of RFAC is shown in Fig. 5.2.

5.3.4 Statistical Analysis

In this phase, two steps are conducted: first, determining the optimal level of each scheduling factor based on S/N ratio results; second, determining the most significant scheduling factors.

The S/N ratio results can be analyzed using statistical tools to determine the optimal conditions for the system's performance. According to Taguchi method, analysis of mean (ANOM) and analysis of variance (ANOVA) can be conducted to predict the optimal factor combinations and identify the most significant factors (Taguchi 1993; Phadke 1989; Roy 2001).

ANOM is used to find the optimal level of each factor. The effect of a factor level is the deviation it causes from the overall mean response. The mean of the S/N ratio (M_{ji}) of factor j at level i can be calculated using the following mathematical expression. The best level of any scheduling factor is the level which gives the highest M_{ji}

$$M_{ji} = \frac{1}{k} \times \sum_{i=1}^{k} \eta_{ji} \qquad (5.10)$$

where

k is the number of levels in factor j

η_{ji} is the value of the S/N ratio with factor j at level i

In some cases, determining the optimal level of factors needs more investigation. In order to do so, a performance measure named the relative percentage deviation (RPD) can be used. The RPD is described using the following equation (Gholami et al. 2009; Naderi et al. 2008).

$$\mathrm{RPD} = \frac{\left[M_{ji} - M_{Min.} \right]}{\left[M_{Min.} \right]} \times 100 \qquad (5.11)$$

where

M_{ji} = the S/N ratio for each instance of factor j at level i.

$M_{Min.}$ − the minimum S/N ratio for each instance of factor j at level i.

ANOVA can be used to determine the percentage contribution of each factor. The largest value of percentage contribution indicates the most significant factor affecting the system performance. The percentage contribution of scheduling factors can be calculated as follows:

1 *Degree of freedom*: The total degree of freedom (df_T), the degree of freedom of factor A (df_A), and the degree of freedom for error variance (df_E) are as follows:

$$df_T = (N - 1) \qquad (5.12)$$

$$df_A = (K_A - 1) \qquad (5.13)$$

$$df_E = \left(df_T - \sum df_{factor} \right) \qquad (5.14)$$

where

N is the total number of experiments.

2 *Sum of squares*: The sum of the square of factor A (SS_A), the total sum of square (SS_T) and the sum of the square for error variance (SS_E) are calculated as follows:

$$SS_A = \sum_{i=1}^{K_A} \left(\frac{A_i^2}{n_{A_i}} \right) - \frac{\left(\sum_{i=1}^{N} x_i \right)^2}{N} \qquad (5.15)$$

$$SS_T = \sum_{i=1}^{N} x_i^2 - \frac{\left(\sum_{i=1}^{N} x_i \right)^2}{N} \qquad (5.16)$$

$$SS_E = \left(SS_T - \sum SS_{factor} \right) \qquad (5.17)$$

where

x_i is a value at level (1, 2, … N).
n_{A_i} is the number of levels and
A_i is a value at level i of factor A.

3 *Mean squares*: The mean square of factor A (MS_A), the total mean square (MS_T) and the mean square of error variance (MS_E) are

$$MS_A = \frac{SS_A}{df_A}, \quad MS_T = \frac{SS_T}{df_T}, \quad MS_E = \frac{SS_E}{df_E} \qquad (5.18)$$

4 *F-ratio*: The value of the *F-ratio* of factor A (F_A) is calculated using the following equation:

$$F_A = \frac{MS_A}{MS_E} \qquad (5.19)$$

5 *Percentage contribution*: the percentage contribution of factor A is calculated using the following equation:

$$PC_A = \frac{SS_A}{SS_T} \times 100 \ \% \qquad (5.20)$$

The following flow chart (Fig. 5.3) summarizes the four phases that are employed to develop a new approach for dynamic scheduling problems in RFAC. This approach will be applied in the next sections using a realistic case study.

5.4   The Simulation Model

The RFAC studied in this model consist of six resources: two robots (R$_1$ and R$_2$) for fetching the assembled parts and placing them at assembly stations (AS$_1$, AS$_2$ and AS$_3$), parts feeder (PF) for supplying parts to the cell, gripper changing station

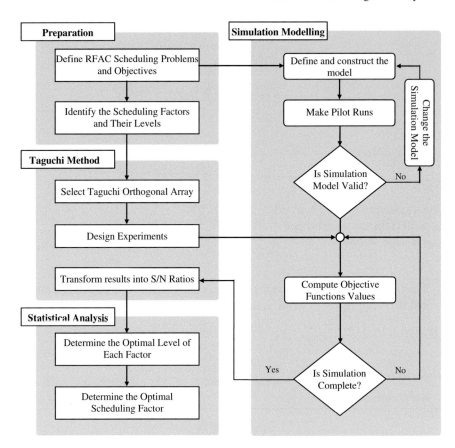

Fig. 5.3 Flow chart of proposed framework

(GC), input conveyors (IC$_1$ and IC$_2$) for supplying the base parts, and output conveyor (OC) for conveying out a final product when assembly processes are completed. Figure 5.4 shows the configuration of the RFAC to be studied.

The presented RFAC model is assumed to assemble n product types ($P_1, P_2, \ldots P_n$). Each product is considered as an independent job. Six products are taken as an example as shown in Table 5.2 and described in Chap. 3. This description shows the details of required stations along with assembly operations time for each product type.

In order to simulate the RFAC model, six customers' orders are considered. These orders are labelled as *Order #1, Order #2… Order #6*, as shown in Table 5.2. Batch size of each product type is also given in Table 5.2.

The RFAC scheduling problem is subject to three classes of constraints, namely robot motion constraints, robot access constraints and tooling resource constraints.

Robot motion constraints: robot arms cannot move from one place to another directly. The reason for this is to avoid collision with the other robot arms. This is

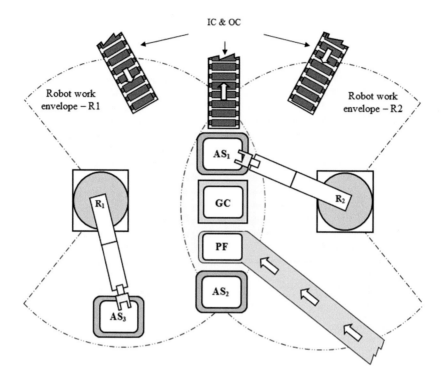

Fig. 5.4 A robotic flexible assembly cell

Table 5.2 Data for customer orders

Product type	Batch size					
	P_1	P_2	P_3	P_4	P_5	P_6
Order$_1$	20	25	25	–	20	25
Order$_2$	30	40	50	40	–	30
Order$_3$	–	25	40	30	20	35
Order$_4$	25	–	20	25	–	20
Order$_5$	30	–	30	30	30	40
Order$_6$	45	20	35	35	50	20

achieved by assigning control points in the cell. Control points $\{C_1, C_2, \ldots, C_4\}$ are set to simplify path planning and avoid collision. For example, R_1 cannot move from AS_1 to PF directly, to achieve this, R_1 should move via control point C_1, as shown in Fig. 5.5.

Robot access constraints: to prevent collisions between robots in a shared area, more than one robot cannot access the same resource simultaneously. For instance, just one robot, R_1 or R_2, can access OC or AS_1 or GC or PF or AS_2 at a time.

Tooling resource constraints: to fetch and assemble parts, the hand of each robot should be equipped with the right tool; however, a specific tool may not be

Fig. 5.5 Robot move
constraints

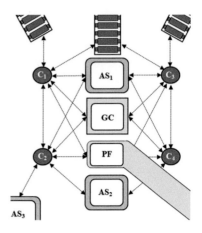

Table 5.3 Transportation time for robots between cell resources

Path description	Position	Travel time (s)
Robot move from resource to control point	S_1, PF $\rightarrow$ C_1, C_3 S_2, S_3, PF $\rightarrow$ C_2, C_4	0.5
Robot move from control point to resource	C_1, C_3 $\rightarrow$ S_1, PF C_2, C_4 $\rightarrow$ S_2, S_3, PF	1
Robot move between control point and conveyor	C_1, C_3 $\leftrightarrow$ IC C_1, C_3 $\leftrightarrow$ OC	1.5
Robot move between two control points	C_1 $\leftrightarrow$ C_2 C_3 $\leftrightarrow$ C_4	0.5

available for the two robots at the same time, due to the restricted number of
available tools.

In this system, four control points $\{C_1, C_2, \ldots, C_4\}$ are set to simplify path
planning and avoid collisions between robots in the shared area. Table 5.3 shows
the robot paths and their required time to move between two positions in the cell.

5.5 Experimental Design and Results

5.5.1 Experimental Setup

As mentioned earlier, the most important scheduling factors, sequencing rules,
dispatching rules, arrival time and due date, are considered in dynamic scheduling
problems of RFAC. To obtain sufficient details of the effect of these factors on
experimental results, each factor is assigned with three levels, because the influence
of these factors may vary nonlinearly. The scheduling factors and the levels for each
factor are briefly described as follows:

Table 5.4 Sequencing rules for the cell to select the order of jobs waiting for processing

Sequencing rule	Description
FCFS	Select the order according to the rule First Come First Served
TSPT	The order with total shortest processing time will be selected
TLPT	The order with total longest processing time will be selected

Table 5.5 Dispatching rules for the robot to select the next job

Dispatching rule	Description
SNQ	Job will be selected according to the smallest number in the queue
SIO	The job with the shortest imminent operation time will be chosen
WINQ	Select the job in the queue which requires the least work

Sequencing rules: When more than one order is waiting for processing, the orders will be sequenced, from the highest order priority to the lowest order priority, using sequencing rules. Three types of sequencing rules are applied (Chan et al. 2002), as shown in Table 5.4.

Dispatching rules: After selecting and loading an appropriate order into the system, the order of jobs has to be dispatched to the available robots. Three types of dispatching rules are used (Chan et al. 2002, 2003) as shown in Table 5.5.

Cell utilisation: In the present research, Eq. 5.1 is used by replacing M with R (number of robots in the system).

$$\propto = \frac{T_p \times N_g}{U \times R} \tag{5.21}$$

Due date tightness: In a dynamic scheduling environment, the total work content (TWK) rule has been extensively used for due date assignment (Ramasesh 1990). Hence, the due date of each job is set using the TWK rule as shown in Eq. 5.4.

Table 5.6 reviews the selected values of the shop utilization and due date tightness factors in recent studies of job and flow shop scheduling problems. Based upon this table, it can be seen that most values of shop utilization factor (U) are between 75 % and 95 %. In the proposed approach, U is fixed at three different values, for example 75, 85, and 95 %. Also, the value of due date tightness factor (K) is between 1 and 10. A high value of K means the required time to finish the job is looser. Conversely, a low value of K means the required time to finish the job is tighter (Lu and Liu 2011). In the proposed example, K is set to three different values, for example 2, 4 and 6.

In this research, the selected factors which influence the scheduling of RFAC, namely sequencing rule (SR), dispatching rule (DR), cell utilization (U) and due date tightness (K), are considered. These factors are set with different levels to explore the effect of the proposed model. The scheduling factors are summarized in Table 5.7.

Table 5.6 Utilization levels and tightness factors reported in recent studies

No.	Author	Year	Shop utilization (%)	Due date tightness
1	Nie et al. (2013)	2013	60, 75 and 90	1, 3 and 5
2	Qiu and Lau (2013)	2013	60, 75 and 90	4, 6 and 8
3	Zhang et al. (2013)	2013	75, 80, 85, 90 and 95	2 and 5
4	Nie (2012)	2012	60 and 90	1, 3 and 5
5	Rajabinasab and Mansour (2011)	2011	85, 90 and 95	4 and 8
6	Zhou et al. (2009)	2009	70, 80 and 90	2
7	Kianfar et al. (2009)	2009	80 and 90	3, 4 and 5
8	Viond and Sridharan (2008)	2008	87 and 90	3, 5 and 7
9	Thiagarajan and Rajendran (2005)	2005	85 and 95	1, 2 and 3
10	Tang et al. (2005)	2005	75, 80, 85 and 90	1.5, 2, 2.5 and 3
11	Dominic et al. (2004)	2004	85 and 95	2, 3, 4, 5
12	Mohanasundaram et al. (2002)	2002	80 and 90	2 and 4

Table 5.7 Scheduling factors and their levels

Factor	Symbol	Factor level	Level ID
Sequencing rules (SR)	A	FCFS	1
		TSPT	2
		TLPT	3
Dispatching rules (DR)	B	SNQ	1
		SIO	2
		WINQ	3
Cell utilisation (U)	C	75 %	1
		85 %	2
		95 %	3
Due date tightness (K)	D	6	1
		4	2
		2	3

5.5.2   Taguchi's Orthogonal Array Selection

Taguchi has developed a pattern of tabulated orthogonal arrays based on the number of factors and their levels. The orthogonal array determines the number of possible experiments. In this investigation, four scheduling factors are taken into consideration with three levels of each factor. Consequently, a total of 81 ($3 \times 3 \times 3$ 3) different combinations are considered. However, according to Taguchi

Table 5.8 Standard L9 (3^4) orthogonal array

Trial No.	Levels of control factors			
	A	B	C	D
1	$A_{(1)}$	$B_{(1)}$	$C_{(1)}$	$D_{(1)}$
2	$A_{(1)}$	$B_{(2)}$	$C_{(2)}$	$D_{(2)}$
3	$A_{(1)}$	$B_{(3)}$	$C_{(3)}$	$D_{(3)}$
4	$A_{(2)}$	$B_{(1)}$	$C_{(2)}$	$D_{(3)}$
5	$A_{(2)}$	$B_{(2)}$	$C_{(3)}$	$D_{(1)}$
6	$A_{(2)}$	$B_{(3)}$	$C_{(1)}$	$D_{(2)}$
7	$A_{(3)}$	$B_{(1)}$	$C_{(3)}$	$D_{(2)}$
8	$A_{(3)}$	$B_{(2)}$	$C_{(1)}$	$D_{(3)}$
9	$A_{(3)}$	$B_{(3)}$	$C_{(2)}$	$D_{(1)}$

method, the number of experiments can be reduced from 81 to 9 using the standard orthogonal array L9 (3^4). Table 5.8 illustrates the arrangement of the experimental design of this study which corresponds to orthogonal array L9.

5.5.3 Calculation of the Signal-to-Noise (S/N) Ratio

The present study uses a robust design criterion called the signal-to-noise (S/N) ratio. The S/N ratio is considered as the key step in Taguchi method and reflects the factor performance. Usually, the S/N ratio characteristics are classified into three types: the *smaller the better*, the *larger the better* and *nominal the best*. Each type calculates the S/N ratio differently. In scheduling problems, nearly all objective functions are classified as the *smaller the better* type; their corresponding S/N ratio is as follows (Phadke 1989):

$$S/_N\text{Ratio} = -10\log(objective\ function)^2 \qquad (5.22)$$

The experimental results of the makespan (C_{max}), the total tardiness (*TD*) and the number of tardy jobs (N_T), with their robust design criteria being based on the combinations of experimental factors, are shown in Tables 5.9, 5.10 and 5.11.

5.6 Analysis of Results and Discussion

5.6.1 Analysis of Mean (ANOM)

The ANOM of the S/N ratio of scheduling factors at any level is calculated using the Eq. 5.10, explained in the Sect. 5.3.4. The final results of ANOM for makespan

Table 5.9 Details of the combinations of different levels and C_{max} results

Trial	Levels of control factors				Makespan	
	A	B	C	D	C_{max}	S/N of C_{max}
1	FCFS	SNQ	Low	Loose	38,499	−91.71
2	FCFS	SIO	Moderate	Moderate	34,736	−90.82
3	FCFS	WINQ	High	Tight	31,571	−89.99
4	TSPT	SNQ	Moderate	Tight	35,387	−90.98
5	TSPT	SIO	High	Loose	32,488	−90.23
6	TSPT	WINQ	Low	Moderate	42,356	−92.54
7	TLPT	SNQ	High	Moderate	33,003	−90.37
8	TLPT	SIO	Low	Tight	39,999	−92.04
9	TLPT	WINQ	Moderate	Loose	36,310	−91.20

Table 5.10 Details of the combinations of different levels and TD results

Trial	Levels of control factors				Total tardiness	
	A	B	C	D	TD	S/N of TD
1	FCFS	SNQ	Low	Loose	10,070	−80.061
2	FCFS	SIO	Moderate	Moderate	19,124	−85.632
3	FCFS	WINQ	High	Tight	50,161	−94.007
4	TSPT	SNQ	Moderate	Tight	28,904	−89.219
5	TSPT	SIO	High	Loose	5666.0	−75.066
6	TSPT	WINQ	Low	Moderate	9909.0	−79.921
7	TLPT	SNQ	High	Moderate	2994.0	−69.525
8	TLPT	SIO	Low	Tight	31,792	−90.046
9	TLPT	WINQ	Moderate	Loose	542.00	−54.680

Table 5.11 Details of the combinations of different levels and N_{TD} results

Trial	Levels of control factors				Number of tardy jobs	
	A	B	C	D	N_{TD}	S/N of N_{TD}
1	FCFS	SNQ	Low	Loose	4	−12.041
2	FCFS	SIO	Moderate	Moderate	12	−21.584
3	FCFS	WINQ	High	Tight	22	−26.848
4	TSPT	SNQ	Moderate	Tight	18	−25.105
5	TSPT	SIO	High	Loose	2	−6.0210
6	TSPT	WINQ	Low	Moderate	7	−16.902
7	TLPT	SNQ	High	Moderate	6	−15.563
8	TLPT	SIO	Low	Tight	19	−25.575
9	TLPT	WINQ	Moderate	Loose	2	−6.0210

Fig. 5.6 Factor responses graph for makespan (C_{max})

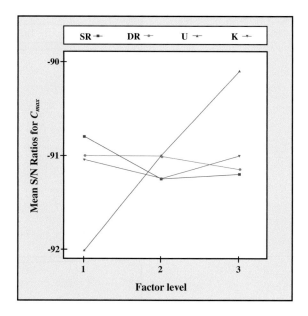

(C_{max}), total tardiness (TD) and number of tardy jobs (N_T) are graphically plotted in Figs. 5.6, 5.7 and 5.8.

Based on the mean S/N ratios presented in Fig. 5.6, the predicted factor level combination that should optimize (i.e., minimize) the makespan (C_{max}) can be seen to be SR_1, DR_1, U_3 and K_3, which may be represented as the sequencing rule (FCFS), the dispatching rule (SNQ), cell utilization = 95 %, and due date tightness = 2.

Regarding the mean S/N ratios presented in Fig. 5.7, the optimal levels of each of the scheduling factors under the measurement of total tardiness (TD) can be seen to be SR_3, DR_3, U_2 and K_1, which may be represented simply as the sequencing rule (TLPT), the dispatching rule (WINQ), cell utilization = 85 %, and due date tightness = 6.

Based on the mean S/N Ratios indicated in Fig. 5.8, the optimal levels of factors U and K under measurement of number of tardy jobs (N_T) are clearly U_3 and K_1. Nevertheless, determining the optimal level of factors SR and DR needs more investigation. In order to do so, the relative percentage deviation (RPD) is used. The RPD results are tabulated in Table 5.12 and plotted in Fig. 5.9.

Regarding the RPD results illustrated in Fig. 5.9, the optimal levels of factors SR and DR are 3 in both cases. Figure 5.9 also demonstrates that the optimal levels of factors U and K are 3 and 1 respectively which confirms and supports the decision on the results presented in Fig. 5.8. Consequently, the optimal levels of scheduling factors under measurement of number of tardy jobs (N_T) can be seen to be SR_3, DR_3, U_3 and K_1, which may be represented easily as the sequencing rule (TLPT), the dispatching rule (WINQ), cell utilization = 95 %, and due date tightness = 6.

Fig. 5.7 Factor responses
graph for total tardiness (*TD*)

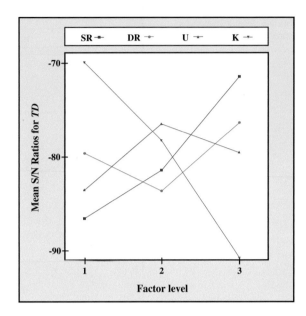

Fig. 5.8 Factor responses
graph for number of tardy
jobs (N_T)

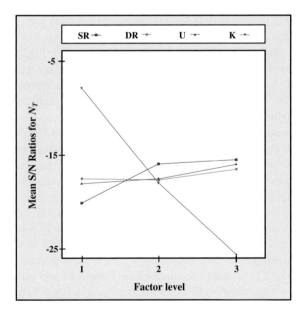

5.6.2 *Analysis of Variance (ANOVA)*

The relative magnitude of the effect of various scheduling factors can be achieved
by employing ANOVA, to determine the percentage of contribution of each factor.

Table 5.12 The mean S/N ratios for expected *TD*

Factor	Mean S/N ratio$_{TD}$	PRD	Factor	Mean S/N ratio$_{TD}$	PRD
SR_1	−20.158	1.511	U_1	−18.173	1.264
SR_2	−16.009	0.994	U_2	−17.570	1.189
SR_3	−15.720	0.958	U_3	−16.144	1.011
DR_1	−17.570	1.189	K_1	−8.0270	0.000
DR_2	−17.726	1.208	K_2	−18.016	1.244
DR_3	−16.590	1.067	K_3	−25.843	2.219

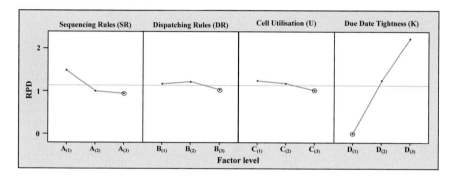

Fig. 5.9 RPD for each level of scheduling factors

The largest value of contribution indicates the most significant factor affecting the
system performance. The procedure of ANOVA consists of four stages to obtain the
contributing scheduling factors: degree of freedom (F), sum of squares (SS), mean
of squares (MS) and percentage of contribution (PC%). All the ANOVA results are
summarized in Tables 5.13, 5.14 and 5.15 where the second column in these Tables
is calculated using Eqs. 5.12–5.14 and the third column is computed according to
Eqs. 5.15–5.17. The fourth and fifth columns are calculated using Eqs. 5.18–5.20
respectively. The following is a sample calculation, using ANOVA equations.

Degree of freedom:

$$df_T = (N - 1) = (9 - 1) = 8$$
$$df_{SR} = (K_{SR} - 1) = (3 - 1) = 2, \quad \text{for factor } (SR)$$
$$df_E = \left(df_T - \sum df_{factor}\right) = (8 - (2 + 2 + 2 + 2)) = 0$$

Table 5.13 ANOVA results for the total tardiness (*TD*)

Factor	F	SS	MS	PC%
SR	2	325.791	162.895	29.94
DR	2	69.361	34.681	6.88
U	2	58.990	29.495	5.92
K	2	645.029	322.515	57.26
Error	0	0.000	0.000	0.00
Total	8	1099.171	137.396	100.00

Table 5.14 ANOVA results for the makespan (*C_{max}*)

Factor	F	SS	MS	PC%
SR	2	0.302	0.151	5.08
DR	2	0.095	0.048	1.60
U	2	5.458	2.729	91.67
K	2	0.098	0.049	1.65
Error	0	0.000	0.000	0.00
Total	8	5.954	0.744	100.00

Table 5.15 ANOVA results for the number of tardy jobs (*N_T*)

Factor	F	SS	MS	PC%
SR	2	64.021	32.011	9.25
DR	2	18.823	9.412	2.72
U	2	11.263	5.632	1.63
K	2	598.059	299.030	86.40
Error	0	0.000	0.000	0.00
Total	8	692.167	86.521	100.00

Sum of squares:

$$SS_T = \left[\left(-80.061^2\right) + \left(-85.632^2\right) + \cdots + \left(-56.521^2\right) \right]$$
$$- \frac{\left[(-80.061) + (-85.632) + \ldots + (-56.521)\right]^2}{9} = 1099.171$$

$$SS_{SR} = \left[\frac{(-80.061 - 85.632 - 94.007)^2}{3} \right] + \left[\frac{(-89.219 - 75.066 - 79.921)^2}{3} \right]$$
$$+ \left[\frac{(-69.525 - 90.046 - 56.521)^2}{3} \right] - \frac{\left[(-80.061) + (-85.632) + \cdots + (-56.521)\right]^2}{9}$$
$$= 325.791$$

$$SS_E = (1099.171 - (325.791 + 69.361 + 58.990 + 645.029)) = 0$$

Mean squares:

$$MS_T = \frac{1099.171}{8}$$

$$MS_{SR} = \frac{325.791}{2} = 162.895$$

Percentage contribution:

$$PC_{SR} = \frac{325.791}{1099.171} \times 100 = 29.64\,\%$$

The result of ANOVA when the makespan (C_{max}) is the objective function is shown in Table 5.13. The table indicates that the cell utilization (U) is the most significant factor affecting the system performance; the contribution of the sequencing rule (SR) is less significant at 5.08 %; the percentage contributions of the other two scheduling factors (K and DR) at 1.65 % and 1.60 % respectively are rather close to each other and so both of these have a less significant influence on performance. The result of ANOVA when the total tardiness (TD) is the objective function is shown in Table 5.14. The table indicates that the due date tightness (K) is the most significant factor affecting the system performance (57.26 %), followed by the sequencing rule (SR) at 29.94 %. The contributions of the dispatching rule (DR) and cell utilization (U) are of low significance at 6.88 % and 5.92 % respectively. The ANOVA, when the objective function is the number of tardy jobs (N_T), is shown in Table 5.15 where the contribution of due date tightness (K) is the most significant at 86.4 %; the contribution of sequencing rule (SR) is less significant at 9.25 %; the contributions of dispatching rule (DR) and cell utilization (U) are of low significance at 2.72 % and 1.63 % respectively.

5.7 Concluding Remarks

The combination of simulation with Taguchi experimental design method offers an efficient and easy to use approach to study the dynamic scheduling problem in RFAC. The proposed approach has many advantageous features: it has the ability to examine the behavior of RFAC under different scheduling factors, to investigate the problem of scheduling with fewer experiments compared to full factorial experimental methods, to find the optimal or near-optimal combination of the selected scheduling factors that optimize the objective functions, and to predict the most significant scheduling factors that have effects on the system performance.

In this chapter, the proposed approach was applied to a scenario-based case study of RFAC. The running time of the scenario is about 12 h. Based on the results and the statistical analysis of these results, the following conclusions can be drawn:

- The factor/level combination $SR_1DR_1U_3K_3$ for the makespan (C_{max}), $SR_3DR_3U_2K_1$ for the total tardiness (TD) and $SR_3DR_3U_3K_1$ for the number of tardy jobs (N_T) are the best recommended parameters for scheduling RFAC when the three objective functions are considered independently.
- The individual factor most affecting the scheduling strategy is cell utilization (U) when considering the C_{max}; and the due date tightness factor (K) is the most significant when the TD and N_T are considered.

Although the proposed approach is designed to deal with dynamic scheduling, it is restricted to solving single-objective optimization problems. This approach can be extended by integrating the Multi-Criteria Decision-Making (MCDM) method with the proposed approach to achieve multi-objective optimization of scheduling problems in RFAC. This has been done by developing a multiple performance characteristics index (MPCI) based on a fuzzy logic approach to derive the optimal solution (Chap. 6).

References

Abd, K., Abhary, K., & Marian, R. (2014a). Simulation modelling and analysis of scheduling in robotic flexible assembly cells using Taguchi method. *International Journal of Production Research, 52*(12), 2654–2666.

Abd, K., Abhary, K., & Marian, R. (2014b). Development of an intelligent approach to dynamic scheduling in robotic fleixble assembly cells. In *IAENG transaction on engineering science* (pp. 203–214). London: Taylor and Francis Group.

Ali, M., & Wadhwa, S. (2010). The effect of routing flexibility on a flexible system of integrated manufacturing. *International Journal of Production Research, 48*(19), 5691–5709.

Bendell, A., Disney, J., & Pridmore, W. A. (1989). *Taguchi methods: Applications in world industry.* Bedford, UK: IFS Publications.

Caprihan, R., & Wadhwa, S. (2005). Scheduling of FMSs with information delays: A simulation study'. *International Journal of Flexible Manufacturing Systems, 17*(1), 39–65.

Caprihan, R., Kumar, A., & Stecke, K. E. (2013). Evaluation of the impact of information delays on flexible manufacturing systems performance in dynamic scheduling environments. *The International Journal of Advanced Manufacturing Technology, 67*(1), 311–338.

Chan, F. T. S., Bhagwat, R., & Wadhwa, S. (2006). Increase in flexibility: Productive or counterproductive? A study on the physical and operating characteristics of a flexible manufacturing system. *International Journal of Production Research, 44*(7), 1431–45.

Chan, F. T. S., Bhagwat, R., & Wadhwa, S. (2007). Flexibility performance: Taguchi's method study of physical system and operating control parameters of FMS. *Robotics and Computer-Integrated Manufacturing, 23*(1), 25–37.

Chan, F. T. S., Bhagwat, R., & Wadhwa, S. (2008). Comparative performance analysis of a flexible manufacturing system (FMS): A review-period-based control. *International Journal of Production Research, 46*(1), 1–24.

Chan, F. T. S., Chan, H. K., & Kazerooni, A. (2003). Real time fuzzy scheduling rules in FMS. *Journal of Intelligent Manufacturing, 14*(3), 341–350.

Chan, F. T. S., Chan, H. K., & Lau, H. C. W. (2002). The state of the art in simulation study on FMS scheduling: A comprehensive survey. *International Journal of Advanced Manufacturing Technology, 19*(11), 830–849.

Cheng, T. C. E., & Gupta, M. C. (1989). Survey of scheduling research involving due date determination decisions. *European Journal of Operational Research, 38*(2), 156–166.

Cowling, P. I., & Johansson, M. (2002). Using real-time information for effective dynamic scheduling. *European Journal of Operational Research, 139*(2), 230–244.

Dominic, P. D. D., Kaliyamoorthy, S., & Kumar, S. M. (2004). Efficient dispatching rules for dynamic job shop scheduling. *International Journal Advance Manufacturing Technology, 24* (1), 70–75.

Gholami, M., Zandieh, M., & Alem-Tabriz, A. (2009). Scheduling hybrid flow shop with sequence-dependent setup times and machines with random breakdowns. *International Journal of Advanced Manufacturing Technology, 42*(1), 189–201.

Jain, A. K., & Elmaraghy, H. A. (1997). Production scheduling rescheduling in flexible manufacturing. *International Journal of Production Research, 35*(1), 281–309.

Kaplan, A. C., & Unal, A. T. (1993). A probabilistic cost-based due date assignment model for job shops. *International Journal of Production Research, 31*(12), 2817–2834.

Kianfar, K., Fatemi Ghomi, S. M. T., & Karimi, B. (2009). New dispatching rules to minimize rejection and tardiness costs in a dynamic flexible flow shop. *International Journal of Advanced Manufacturing Technology, 45*(7), 759–771.

Lee, H. H. (2000). *Taguchi methods: Principles and practices of quality design.* Taiwan: Gaulih Book Co. Ltd.

Lu, M. S., & Liu, Y. J. (2011). Dynamic dispatching for a flexible manufacturing system based on fuzzy logic. *The International Journal of Advanced Manufacturing Technology, 54*(9–12), 1057–1065.

Mohanasundaram, K. M., Natarajan, K., Viswanathkumar, G., Radhakrishnana, P., & Rajendran, C. (2002). Scheduling rules for dynamic shops that manufacture multi-level jobs. *Computers & Industrial Engineering, 44*(1), 119–131.

Moradi, E., Ghomi, S. M. T. F., & Zandieh, M. (2010). An efficient architecture for scheduling flexible job-shop with machine availability constraints. *The International Journal of Advanced Manufacturing Technology, 51*(1–4), 325–339.

Naderi, B., Zandieh, M., & Roshanaei, V. (2008). Scheduling hybrid flowshops with sequence dependent setup times to minimize makespan and maximum tardiness. *International Journal of Advanced Manufacturing Technology, 41*(11), 1186–1198.

Nie, L. (2012). Discover scheduling strategies with gene expression programming for dynamic flexible job shop scheduling problem. *Advances in Swarm Intelligence, 7332*, 383–390.

Nie, L., Gao, L., Li, P., & Li, X. (2013). A GEP-based reactive scheduling policies constructing approach for dynamic flexible job shop scheduling problem with job release dates. *Journal of Intelligent Manufacturing, 4*(4), 763–774.

Ouelhadj, D., & Petrovic, S. (2009). A survey of dynamic scheduling in manufacturing systems. *Journal of Scheduling, 12*(4), 417–31.

Phadke, M. S. (1989). *Quality engineering using robust design.* USA: Prentice-Hall.

Qiu, X., & Lau, H. Y. K. (2013). An AIS-based hybrid algorithm with PDRs for multi-objective dynamic online job shop scheduling problem. *Applied Soft Computing, 13*(3), 1340–1351.

Rajabinasab, A., & Mansour, S. (2011). Dynamic flexible job shop scheduling with alternative process plans: an agent-based approach. *International Journal of Advanced Manufacturing Technology, 54*(9–12), 1091–1107.

Ramesesh, R. (1990). Dynamic job shop scheduling: A survey of simulation research. *OMEGA: International Journal of Management Science, 18*(1), 43–57.

ROSS, P. J. (1988). *Taguchi techniques for quality engineering.* New York: McGraw-Hill.

Roy, R. K. (2001). Design of experiment using the Taguchi approach: 16 Steps to product and process improvement. New York: John Wiley & Sons Inc.

Stoop, P. P. M., & Weirs, V. C. S. (1996). The complexity of scheduling in practice. *International Journal of Operations and Production Management, 16*(10), 37–53.

Taguchi, G. (1993). *Taguchi on robust technology development.* New York: ASME Press.

Taguchi, G., Elsayed, E., & Hsaing, T. (1989). Quality engineering in production systems. McGraw-Hill, New York.

Tang, L., Liu, W., & Liu, J. (2005). A neural network model and algorithm for the hybrid flow shop scheduling problem in a dynamic environment. *Journal of Intelligent Manufacturing, 16* (3), 361–370.

Thiagarajan, S., & Rajendran, C. (2005). Scheduling in dynamic assembly job-shops to minimize the sum of weighted earliness, weighted tardiness and weighted flowtime of jobs. *Computers and Industrial Engineering, 49*(4), 463–503.

Vieira, G. E., Herrmann, J. W., & Lin, E. (2003). Rescheduling manufacturing systems: A framework of strategies, policies and methods. *Journal of Scheduling, 6*(1), 36–92.

Vinod, V., & Sridharan, R. (2008). Scheduling a dynamic job shop production system with sequence-dependent setups: An experimental study. *Robotics and Computer-Integrated Manufacturing, 24*(3), 435–449.

Xiang, W., & Lee, H. P. (2008). Ant colony intelligence in multi-agent dynamic manufacturing scheduling. *Engineering Applications of Artificial Intelligence, 21*(1), 73–85.

Zhang, L., Li, X., Gao, L., & Zhang, G. (2013). Dynamic rescheduling in FMS that is simultaneously considering energy consumption and schedule efficiency. *International Journal of Advanced Manufacturing Technology,*. doi:10.1007/s00170-013-4867-3.

Zhou, R., Nee, A. Y. C., & Lee, H. P. (2009). Performance of an ant colony optimisation algorithm in dynamic job shop scheduling problems. *International Journal of Production Research, 47* (11), 2903–2920.

Chapter 6
Development of an Optimization Approach for Dynamic Scheduling Problems in RFAC

6.1 Introduction

In Chap. 5 the proposed approach designed to solve *single-objective* optimization problems for dynamic scheduling was described. The aim of this chapter is to extend the approach proposed in Chap. 5 to solving *multi-objective* optimization problems. Some of the research studies in this field have attempted to solve dynamic scheduling problems which have multi-objectives. The methodologies described in these studies have been achieved through simulation or artificial intelligence approaches; however, there is a lack of methodologies for dealing with optimization problems. This deficiency is due to the weighting method (i.e. assigning weights for each objective) used to transform a multi-objective problem to a single-objective problem and optimize the new single-objective problem e.g. Adibi et al. 2010; Fattahi and Fallahi 2010; Tavakkoli-Moghaddam et al. 2011a, b; Zhang et al. 2013). Commonly, the weights are determined by schedulers based on their own judgments of the importance of each objective (Rangsaritratsamee et al. 2004; Xing et al. 2009). In fact, the weighting method is based purely on subjective assessment by human experts. Consequently, this method still cannot reflect the human thinking process. In order to deal with the subjectivity of human thought, fuzzy set theory can be used. The main feature of fuzzy sets is their ability to mathematically represent vague knowledge (Beskese et al. 2004; Azadegan et al. 2011).

One class of fuzzy set methodologies' *Fuzzy Multi-Criteria Decision-Making* (Fuzzy MCDM), considers optimization problems that may have more than one objective function (Kahraman et al. 2010; Abdullah 2013). The objective of this part of the research is to solve multi-objective optimization problems for dynamic scheduling in RFAC, using fuzzy MCDM. In order to achieve this objective, the following two steps are accomplished.

- Review of the relevant literature for the use of the most well-known MCDM methods, and then focus on the recent methods that have been used to deal with imprecise and uncertain information.

© Springer International Publishing Switzerland 2016
K.K. Abd, *Intelligent Scheduling of Robotic Flexible Assembly Cells*,
Springer Theses, DOI 10.1007/978-3-319-26296-3_6

- Development of a new fuzzy MCDM approach to solve multi-objective optimization problems for dynamic scheduling in RFAC.

The approach developed in this chapter will then be examined, verified and validated in Chap. 7, using a realistic case study.

6.2 Multi-criteria Decision-Making

Decision-making is the process of evaluating a number of feasible solutions and finding the optimal one. Sometimes, decision-making problems dealing with several criteria with different effects are called multi-criteria decision-making (MCDM) problems (Korhonen et al. 1992; Figueira et al. 2005).

In the literature, other acronyms are also used for MCDM which essentially have the same meaning: for example, multi criteria decision analysis (MCDA), multi objective decision support (MODS), and multi attribute decision making (MADM) (Hyde 2006). In this thesis, the acronym MCDM will be used. MCDM has three components: first, a finite number of *alternatives* (solutions); second, a set of *criteria* by which the alternatives are to be judged; and third, a method for evaluating and ranking the alternatives based on how well they satisfy the criteria (Figueira et al. 2005). The terms *alternatives* and *criteria* are defined as follows:

- Alternatives represent the number of solutions available to the decision maker.
- Criteria represent different dimensions of the alternatives. These criteria may be in conflict with each other.

6.2.1 MCDM Process

The MCDM process refers to the procedure of problem solving. Generally, the process of MCDM involves five stages: (1) defining problems to be examined and the objectives to be optimized; (2) identifying the possible alternatives, and the multiple criteria; (3) selecting the most appropriate MCDM method; (4) aggregating the outcomes of the MCDM method and combining them into one rank order; and (5) ranking the alternatives from the optimal to the least desirable one (Triantaphyllou 2000; Brugha 2004). Figure 6.1 shows the stages of MCDM process.

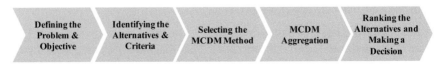

Fig. 6.1 Summary of MCDM process

6.2.2   In Search of a Powerful Method for Complex Decision Making Problems

MCDM method is extensively used for evaluating a number of feasible solutions against a number of criteria and determining the best solution which meets all the criteria. Selecting a suitable MCDM method is partly based on the decision maker's preference. In recent years, many methods have been proposed for solving MCDM problems. Examples of these methods are WSM (Weighted Sum Model), WPM (Weighted Product Model), AHP (Analytic Hierarchy Process), TOPSIS (Technique for Order Preference by Similarity to Ideal Solution) and SMART (Simple Multi-Attribute Rating Technique) (Triantaphyllou 2000; Guarnieri 2014). There is no universally accepted MCDM method, but some methods are more suitable than others for solving particular decision problems (Dagdeviren 2008). AHP and TOPSIS are the most well-known MCDM methods (Aydogan 2011; Samvedi et al. 2013).

AHP method was originally developed by Saaty (1980). The AHP method is widely used and can effectively handle both quantitative and qualitative data in different practical decision making problems (Kahraman et al. 2004; Javanbarg et al. 2012). This method consists of three main stages: first, constructing a pair-wise comparison matrix; second, synthesizing judgments; third, checking for consistency (Vaidya and Kumar 2006; Amiri 2010; Abd et al. 2011).

Even though AHP has been successfully applied to solve different decision problems, it has a number of drawbacks. First, this method is still used in nearly crisp decision applications (Patil and Kant 2014; Ayag 2005). Second, AHP cannot reflect the human thinking style, due to the fact that it deals with subjective assessments (Wang et al. 2008; Patil and Kant 2014; Seçme et al. 2009). Third, it does not have the ability to handle decision problems when the information is uncertain and ambiguous (Torfi et al. 2010; Zouggari and Benyoucef 2012).

To overcome the above drawbacks, AHP is combined with fuzzy set theory, the so called fuzzy AHP method (FAHP) widely used by researchers, mainly in engineering and technology applications, including for evaluation of computer integrated manufacturing systems (Bozdağ et al. 2003), software project selection (Buyukozkan et al. 2004), operating system selection (Tolga et al. 2005), evaluation of production planning and scheduling (Kang and Lee 2007), selection of plant location (Kaboli et al. 2007), supplier selection (Chan et al. 2008; Chamodrakas et al. 2010), machine tool selection (Duran and Aguilo 2008), course website quality evaluation (Lin 2010).

TOPSIS was initially presented by Hwang and Yoon (1981). TOPSIS is one of the most common methods used to solve MCDM problems. The basic idea of TOPSIS is that the selected alternative should be close to the best of ideal solution and farthest from the worst one (Lin 2008; Shyjith 2008). The features of this method are that it is easy to use, the process is understandable, and also it is capable of taking into account both quantitative and qualitative criteria (Ekmekçioğlu et al. 2010). Like AHP, TOPSIS has the same drawback of not being suitable for making decisions when the information is uncertain. Therefore, TOPSIS is combined with fuzzy set theory, the so called fuzzy TOPSIS (FTOPSIS), extensively applied in real case studies to deal with MCDM problems. Examples of FTOPSIS applications are robot selection (Chu and Lin 2003), plant location selection (Yong 2006), supplier selection (Boran et al. 2009) and evaluation of manufacturing plants (Yu and Hu 2010).

Other studies, as shown in Table 6.1, integrated FAHP with TOPSIS or AHP with FTOPSIS for evaluating decision problems under different criteria. In these studies, the AHP/FAHP is used to determine the relative weights of the multiple evaluation criteria, and the TOPSIS/FTOPSIS is used for evaluation of each alternative in order to make a final decision. Although these studies indicated that the integration of FAHP with TOPSIS or AHP with FTOPSIS are suitable approaches for solving different decision problems in engineering and technology, they are unable to fully handle the uncertainty and imprecision in real world problems. Thus, the recent studies have integrated FAHP with FTOPSIS; this combination can adequately handle decision problems containing uncertain and ambiguous information, and can also provide the flexibility needed for the decision maker to understand the decision problem. Table 6.1 lists the recent studies which integrate FAHP with FTOPSIS.

In recent decades, several studies in the literature have developed a fuzzy decision support system (FDSS) as an MCDM tool, to solve real-world problems (Abd et al. 2013a). In these studies, the FDSS is well implemented in different application areas, as shown in Table 6.1, in order to deal with vagueness and uncertainty. These studies fully utilize the capability of the Matlab fuzzy logic toolbox to build a computer-based system for supporting complex decision making and problem solving.

Based on the previous literature, it can be stated that the FDSS and FAHP-FTOPSIS are efficient and modern approaches for solving MCDM problems. However, to the best of the author's knowledge, there is no work published that uses fuzzy MCDM method for optimization of dynamic scheduling problems. Moreover, there is no documented research in the area of MCDM that indicates the application of both FDSS and FAHP-FTOPSIS approaches. This chapter attempts to develop a new approach based on FDSS and FAHP-FTOPSIS for solving optimization problems. The main advantages of the proposed approach are: (1) it can mimic human expert reasoning for optimizing the dynamic scheduling in RFAC (2) it can effectively handle the problems when the information is uncertain and ambiguous (3) it is powerful enough to deal with complex decision making problems. The next section will describe this approach in detail.

Table 6.1 Fuzzy MCDM approaches and their applications

Approach	Author	Year	Specific area
Fuzzy AHP-TOPSIS	Seçme et al.	2009	Evaluation of banks' performances
	Gumus	2009	Selection of transportation firm
	Wang et al.	2010	Aeroengine health assessment
	Azadeh et al.	2011	Evaluation of cellular manufacturing systems
	Jafarian and Vahdat	2012	Welding process selection
	Kumar and Singh	2012	Supply chain problem
	Parsaei et al.	2012	Selection of the best order
	Choudhary and Shankar	2012	Location selection for power plant
AHP-fuzzy TOPSIS	Önüt and Soner	2008	Site selection
	Dagdeviren et al.	2009	Weapon selection
	Amiri	2010	Project selection for oil-fields
	Yu et al.	2011	Website evaluation
	Aydogan	2011	Evaluation of manufacturing system performance
	Awasthi and Chauhan	2012	Evaluating city logistics initiatives
	Maldonado-Macías et al.	2014	Evaluation of CNC milling machines
Fuzzy AHP-fuzzy TOPSIS	Sun	2010	Evaluation of computer companies
	Rostamzadeh and Sofian	2011	Production systems performance
	Buyukozkan and Cifci	2012	Electronic service quality performance
	Mentes and Helvacioglu	2012	Mooring system selection
	Zouggari and Benyoucef	2012	Global supplier selection
	Paksoy et al.	2012	Determinants of distribution channel management
	Kutlu and Ekmekcioglu	2012	Evaluation of failure mode and effects analysis
	Samvedi et al.	2013	Evaluation of supply chain risks
	Aktan and Tosun	2013	Automated storage and retrieval system selection
	Patil and Kant	2014	Supply chain problem
Fuzzy decision support system	Haji and Assadi	2009	Determining new product pricing
	Fasanghari and Montazer	2010	Stocks selection
	Damghani et al.	2011	Optimal project selection
	Berihaa et al.	2012	Evaluation of safety performance in industries

(continued)

Table 6.1 (continued)

Approach	Author	Year	Specific area
	Hesami et al.		Determining adjusted price of products
	García et al.	2013	Supplier selection
	Bugarski et al.	2013	Evaluation of ship lock control
	Camastra et al.	2014	Environmental risk assessment

6.3   A Hybrid Approach for Optimization of Dynamic Scheduling Problems

In this chapter a comprehensive hybrid fuzzy MCDM is developed using FDSS and FAHP-FTOPSIS. The aim of the proposed approach is to optimize *multi-objective* dynamic scheduling problems in RFAC. To address the above aim, the fuzzy MCDM approach is divided into three phases: problem description; application of fuzzy MCDM; and analysis of the results, as shown in Fig. 6.2. Each of the phases is described in the following sub-sections.

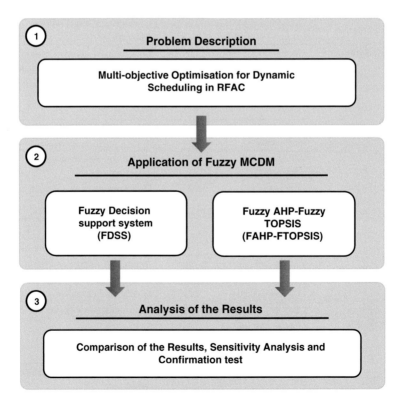

Fig. 6.2 Simplified flow chart of the developed approach

6.3.1   Problem Description

In this phase, the MCDM problem must be described as a hierarchy. The hierarchical structure of the problem can be arranged based on the overall goal, criteria and alternatives (Triantaphyllou 2000). Four steps are required to form the hierarchy of any MCDM problem: (1) define the research problem and identify the overall goal; (2) determine the criteria and sub-criteria that must be used to fulfill the overall goal; (3) identify the decision alternatives; and (4) structure the hierarchy by placing the objective at the top level, the criteria and sub-criteria in the middle and the decision alternatives at the bottom level. The whole hierarchy of the decision making problem can be easily drawn as shown in Fig. 6.3.

Once the hierarchical structure is established, a decision table is built. In this table, the columns indicate the criteria (objective functions), and the rows indicate the alternatives (feasible solutions). The intersections between columns and rows represent the cells. The value in each cell represents the evaluation of the feasible solution with respect to the single-objective. The value range for the multi-objective is generally different. To avoid the different ranges, the values in each column must be normalized to values from 0 to 1. There are two scenarios to perform normalization:

- In the first scenario, the objective value is of the *smaller the better* type. In this case, the normalization can be determined using Eq. (6.1).
- In the second scenario, the objective value is of the *larger the better* type. In this case, the normalization is determined using Eq. (6.2).

$$\mu_k^i = \frac{\left[\eta_{i(k)} - \text{Min } \eta_{i(k)}\right]}{\left[\text{Max } \eta_{i(k)} - \text{Min } \eta_{i(k)}\right]}, \quad 0 \leq \mu_k^i \leq 1 \tag{6.1}$$

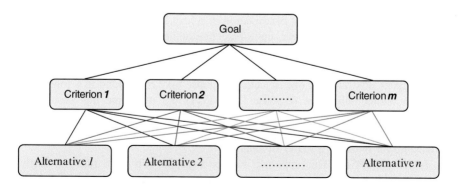

Fig. 6.3 The hierarchy of the MCDM problem

$$\mu_k^i = \frac{\left[\text{Max } \eta_{i(k)} - \eta_{i(k)}\right]}{\left[\text{Max } \eta_{i(k)} - \text{Min } \eta_{i(k)}\right]}, \quad 0 \le \mu_k^i \le 1 \tag{6.2}$$

where μ_k^i is the normalised objective value for the kth response in the ith feasible solution; $\eta_{i(k)}$ is the value of the objective; Min $\eta_{i(k)}$ and Max $\eta_{i(k)}$ are the smallest and the largest value of $\eta_{i(k)}$ respectively.

6.3.2 Application of Fuzzy MCDM Methods

MCDM methods under the fuzzy environment are applied in order to deal with the vagueness of human thought in making decisions. In this phase, three fuzzy MCDMs will be used due to their unique characteristics as stated previously. These fuzzy MCDMs are:

- Fuzzy decision support system (FDSS)
- Fuzzy AHP (FAHP)
- Fuzzy TOPSIS (FTOPSIS)

FDSS: The motivations for developing a FDSS can be summarized as follows: (1) it mimics human expert reasoning for solving real decision problems (García et al. 2013; Fasanghari and Montazer 2010); (2) it has the ability to deal with imprecise and uncertain information (Damghani et al. 2011; Haji and Assadi 2009); (3) it is more intuitive in its ability to include knowledge in the decision making process and enhance its interpretation (Berihaa et al. 2012; Fasanghari and Montazer 2010); and (4) the computation processes are faster than other decision methods due to their intrinsic parallel processing nature (García et al. 2013). In this part of the research, a FDSS is developed in order to solve multi-objective optimization problems for dynamic scheduling in RFAC.

As mentioned earlier, the proposed approach in Chap. 5 is designed to solve single-objective optimization problems for dynamic scheduling in RFAC. For a single-objective, the optimum combination of scheduling factors is the one having the highest S/N ratio. However, multi-objective optimization is not as straightforward as single-objective optimization. As a result, an overall evaluation of S/N ratios is required for the optimization of the multi-objective response due to the fact that a higher S/N ratio for one objective response may correspond to a lower S/N ratio for another objective response (Rao 2011). To overcome this problem, a multiple performance characteristics index (MPCI) based on FDSS is calculated in order to derive the optimal solution. Section 6.4 will describe the methodology and procedure for building a FDSS in detail.

FAHP-FTOPSIS: This approach will be used for two main reasons: firstly, recent studies have showed that the combination of fuzzy AHP method (FAHP) with fuzzy TOPSIS method (FTOPSIS) provides a powerful approach to deal with complex decision problems; secondly, this approach has the ability to handle decision problems when the information is uncertain due to vagueness and imprecision (Patil and Kant 2014; Aktan and Tosun 2013; Zouggari and Benyoucef 2012).

In this research, the FAHP is combined with FTOPSIS for optimizing the dynamic scheduling in RFAC under different performance measures. The FAHP is applied to determine the weights of the multiple evaluation criteria to be used in the evaluation process; and the FTOPSIS is applied for evaluation of each alternative scheduling combination based on its overall performance in order to make a final decision. Section 6.5 will illustrate in detail the methodology and algorithm for combining FAHP with FTOPSIS in detail.

6.3.3   Analysis of the Results

In this phase, the outcome results of FDSS and FAHP-FTOPSIS are analyzed and discussed. This phase includes four steps: first, comparing the results of FDSS with those of FAHP-FTOPSIS; second, checking the stability of the FDSS and FAHP-FTOPSIS results by applying a *sensitivity analysis*; third, determining the *effect of scheduling factors* on system performance; fourth, verifying the final results via a *confirmation test*. The sensitivity analysis, effect of scheduling factors and confirmation test are explained below:

Sensitivity analysis: A sensitivity analysis is conducted to analyze the obtained results of FMCDM and Fuzzy AHP-Fuzzy TOPSIS. The purpose of this analysis is to investigate the influence of the criteria weights on the final performance ranking, and to check whether a few changes in the criteria weights lead to significant modification in the decision outcome. Therefore, sensitivity analysis provides information on the stability of the results. Based on the previous studies in MCDM, there are two procedures to achieve sensitivity analysis. The first procedure is to increase or decrease the weights of each individual criterion, and keep the summation total of all criteria weights equal to one (Kutlu and Ekmekcioglu 2012). The other procedure is to exchange each criterion's weight with another criterion's weight while the other remaining criteria are constant (Mentes and Helvacioglu 2012; Azadeh et al. 2011).

Effect of scheduling factors on MPCI: The effect of scheduling factors based on S/N ratio results can be determined. In this step, the significant factor combinations in terms of their contribution to the multi-objective functions are indicated. This can be calculated from the mean of each factor level, using the following mathematical expression (Phadke 1989):

$$MPCI_{ji} = \frac{1}{k} \times \sum_{i=1}^{k} \eta_{ji} \qquad (6.3)$$

where $MPCI_{ji}$ is the mean of factor j at level i; k is the number of levels in factor j; and η_{ji} is the value of MPCI with factor j at level i. The highest value of MPCI among all combinations of the factors denotes the optimum level for each factor.

Confirmation test: The final step is to predict and verify the improvement of the performance characteristics using the optimal levels of the scheduling factors. The best value of the MPCI can be predicted using the following Equation (Montgomery 2004):

$$\eta_{MPCI} = \eta_m + \sum_{i=1}^{q}(\eta_i - \eta_m) \tag{6.4}$$

where η_m is the total mean of the MPCI, η_i is the mean of the MPCI at the optimal level and q is the number of scheduling factors.

In addition, the percentage prediction error (PPE) can be calculated in order to check the closeness of the predicted value to that obtained by the actual experimental value (Pandey and Dubey 2012).

$$PPE = \frac{[|Experimental\ value - Predicted\ value|]}{(Experimental\ value)} \times 100 \tag{6.5}$$

6.4   Implementation of Fuzzy Decision Support System

In this section, the structure and design of the proposed FDSS is demonstrated and explained. The FDSS will be used to calculate a multiple performance characteristics index (MPCI), to find the optimal solutions of multi-objective problems for dynamic scheduling in RFAC.

6.4.1   Structure of Fuzzy Decision Support System

A FDSS consists of four components: fuzzification, decision rules, knowledge base and defuzzification (Abd et al. 2013b), as shown in Fig. 6.4. These components are explained in detail below:

Fuzzification: Fuzzification is the process of transforming the real world variables into linguistic variables. The fuzzification process can be achieved using the membership functions of input variables. This process involves three stages: measuring the values of crisp input data; performing a scale mapping that converts the range of input variables' values into a corresponding normalized universe of discourse; performing the function of fuzzification operator $x = fuzzifier\,(x_0)$, which has the effect of converting crisp input data into a fuzzy set, where

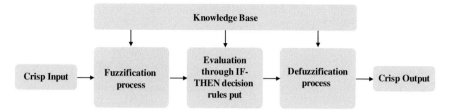

Fig. 6.4 The architecture of the FDSS

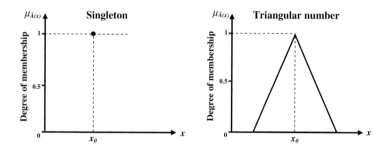

Fig. 6.5 Two examples of fuzzification function

x represents a fuzzy set, fuzzifier denotes a fuzzification operator, and x_0 represents crisp data (Deb and Bhattacharyya 2005; Wang 1997) (Fig. 6.5).

Decision rule: The decision rule is the most important component in FDSS where fuzzy input values map to fuzzy output based on expert knowledge. The expert knowledge is usually described in the form of IF-THEN fuzzy rules. The number of fuzzy rules depends on the number of input variables and their linguistic values. The fuzzy decision rule is expressed as follows:

$$Rule\ i = \text{IF } X_1 \text{ is } A_1^i \text{ and} \dots \text{ and } X_n \text{ is } A_n^i \text{ THEN } Y \text{ is } B^i, \quad i = 1, 2, \dots, M \quad (6.6)$$

where $X = (X_1, X_2, \dots, X_n)$ and Y are the input and output linguistic fuzzy variables of the fuzzy system, respectively. $A = (A_1, A_2, \dots, A_n)$ and B are fuzzy sets in a predefined universe of discourse. M is the number of fuzzy rules.

Knowledge base: The knowledge base involves the necessary information that is required for the fuzzification interface, the decision rules interface and the defuzzification interface. This information contains: (1) the fuzzy rule base that is usually described by the expert knowledge and (2) the membership functions for the fuzzy set that is constructed to perform fuzzification and defuzzification processes. The shape of the membership functions can be defined by specifying their geometry value numbers. For example, the triangular fuzzy numbers (TFN) can be defined by a triplet $l \leq m \leq n$, where l, m and n are the lowest possible value, the most credible

Fig. 6.6 Defuzzification
methods for obtaining the
crisp output

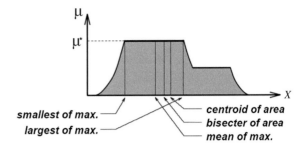

value, and the highest possible value respectively. Then, the TFN can be defined by its continuous membership function $\mu_{\tilde{A}}(x) : X \rightarrow [0, 1]$ as follows:

$$\mu_{\tilde{A}}(x) = \begin{cases} (x - l)/(m - l), & l \leq x \leq m \\ (n - x)/(n - m), & m \leq x \leq n \\ 0, & \textit{otherwise} \end{cases} \tag{6.7}$$

A trapezoidal fuzzy numbers (TrFN), as another example, is represented by four points $l \leq m \leq n \leq P$. The TrFN can be defined by its continuous membership function $\mu_{\tilde{A}}(x) : X \rightarrow [0, 1]$ given by:

$$\mu_{\tilde{A}}(x) = \begin{cases} (x - l)/(m - l), & l \leq x \leq m \\ 1, & m \leq x \leq n \\ (p - x)/(p - n), & n < x \leq p \\ 0, & \textit{otherwise} \end{cases} \tag{6.8}$$

Defuzzification: Defuzzification is the process of converting a fuzzy set into a crisp value. The defuzzification process can be achieved using the membership functions of the output variables. In general, there are five defuzzification methods in the literature: *center of gravity* (COA), *mean of maximum* (MOM), *smallest of maximum* (SOM), *largest of maximum* (LOM) and *bisector of area* (BOA). Figure 6.6 shows the defuzzification methods for obtaining the crisp output. The definitions of popular defuzzification methods are shown below (Lee 2005):

Centroid of area (COA): the COA is often referred to as the center of gravity because it computes the centroid of the composite area representing the output fuzzy term. The crisp output X_{COA} given the membership $\mu_{\tilde{A}}(x)$ is computed as:

$$X_{COA} = \frac{\int \mu_{\tilde{A}}(x) x \, dx}{\int \mu_{\tilde{A}}(x) \, dx} \tag{6.9}$$

Bisector of area (BOA): the bisector is the vertical line that divides the area into two regions with the same area. The X_{BOA} is computed as follows, where $\alpha = \min (x | x \in X)$ and $\beta = \max (x | x \in X)$.

$$\int\limits_{\propto}^{X_{BOA}} \mu_{\tilde{A}}(x)\,dx = \int\limits_{X_{BOA}}^{\beta} \mu_{\tilde{A}}(x)dx \qquad (6.10)$$

Mean of maximum (MOM): the MOM is the average of the maximizing X at which the membership function reaches a maximum μ^*. The crisp output X_{MOM} is computed as follows, where $x' = (x \mid \mu A(x) = \mu^*)$.

$$X_{MOM} = \frac{\int x'xdx}{\int x'dx} \qquad (6.11)$$

6.4.2   Design of the Proposed Fuzzy Decision Support System

In this section, a FDSS is proposed to optimize the dynamic scheduling problems in RFAC. The FDSS methodology can be divided into four steps: first, defining the input-output variables; second, specifying the membership functions of all the system variables; third, building fuzzy decision rules of the proposed system; and fourth, determining the final output (*MPCI*). The four steps are explained in detail below.

Step 1: Defining input and output variables
The first step is to define the input and output variables of the proposed FDSS. The input variables represent the S/N ratio of each objective function individually. The output variable represents the *MPCI*. Figure 6.7 shows the input and output block diagram of the proposed FDSS.

Step 2: Specifying input and output membership functions
After identifying inputs/output parameters, the membership functions for these parameters are specified. There are different shapes of membership functions such as triangular, trapezoidal, Gaussian, singleton, etc. The triangular fuzzy numbers (TFN) and trapezoidal fuzzy numbers (TrFN) are the common membership functions shapes and a powerful way to approach the convex function. Thus, the TFN and TrFN are used in this research to construct the input and output variables of the

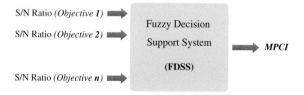

Fig. 6.7 Block diagram of a FDSS with input and output variables

Fig. 6.8 Membership functions of linguistic output variables

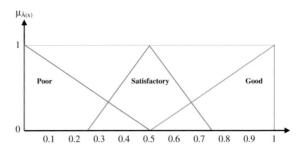

Fig. 6.9 Membership functions of linguistic output variables

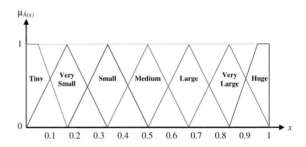

proposed system. The membership functions of the S/N ratios are broken into a set of linguistic values for the inputs: poor (0, 0, 0.5), satisfactory (0.25, 0.5, 0.75), and good (0.5, 0.75, 1), while the membership functions of the MPCI are set into seven linguistic values: tiny (0, 0, 0.05, 0.166), very small (0, 0.166, 0.33), small (0.166, 0.33, 0.50), medium (0.33, 0.50, 0.66), large (0.50, 0.66, 0.83), very large (0.66, 0.83, 1) and huge (0.83, 0.95, 1, 1). The input variables (Fig. 6.8) are constructed as a triangular shape. The output variable (MPCI) is built from both triangular and trapezoidal shapes, as depicted in Fig. 6.9.

Step 3: Constructing decision rules and knowledge base

The output variable (MPCI) is controlled using fuzzy rules. The number of fuzzy rules is derived directly based on the formula (n^m), where n and m denote input variables (S/N Ratios) and their linguistic values respectively. The general form of a fuzzy IF – THEN rule for the proposed FDSS can be expressed using Eq. (6.6). Examples of the fuzzy rules derived are shown below:

Rule 1 = IF S/N *Ratio*$_1$ *is Poor and* … *and S/N Ratio*$_n$ *is Poor* THEN *MPCI is* …

Rule 2 = IF S/N *Ratio*$_1$ *is Poor and* … *and S/N Ratio*$_n$ *is Satisfactory* THEN *MPCI is* …

Rule 3 = IF S/N *Ratio*$_1$ *is Poor and* … *and S/N Ratio*$_n$ *is Good* THEN *MPCI is*. …

$\vdots$

Rule M = IF S/N *Ratio*$_1$ *is Good and* … *and S/N Ratio*$_n$ *is Good* THEN *MPCI is* …

Step 4: Determining MPCI by using defuzzification

The evaluation values of each fuzzy rule are combined into one single output value via the defuzzification process. There is still no systematic method to achieve the defuzzification process. In this research, the well-known method for the defuzzification process which is called the center of area (*COA*) is used. The *COA* transforms the fuzzy inference output $\mu\tilde{A}$ (x) into the non-fuzzy value x'. The non-fuzzy value x' is called MPCI, based on the following equation:

$$MPCI = \frac{\int \mu\tilde{A}(x)\, x\, dx}{\int \mu\tilde{A}(x)\, dx} \qquad (6.12)$$

After achieving the above four steps, the proposed system is ready to be applied to realistic problems for optimizing the dynamic scheduling in RFAC under different performance measures. The configuration of the proposed FDSS is depicted in Fig. 6.10. This figure shows the FDSS components: fuzzification, decision rules, knowledge base and defuzzification:

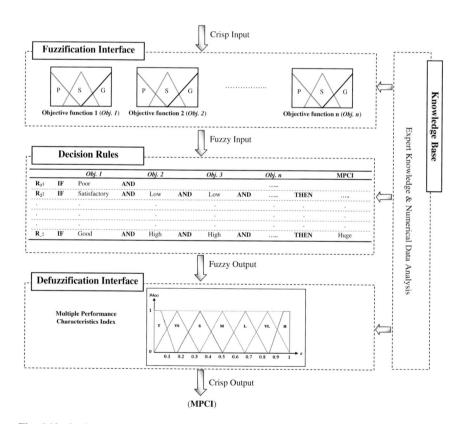

Fig. 6.10 Outline of the developed fuzzy decision support system

6.5 Implementation of Fuzzy AHP-Fuzzy TOPSIS

In this section, fuzzy AHP (FAHP) integrated with fuzzy TOPSIS (FTOPSIS) for optimizing the dynamic scheduling in RFAC under different performance measures is proposed as a new application. The FAHP is used to determine the relative weight of the multiple evaluation criteria, and the FTOPSIS is applied for evaluation of each alternative in order to make a final decision. The methodology and algorithm steps of FAHP and FTOPSIS are summarized as follows:

6.5.1 Methodology of FAHP

FAHP consists of three main stages: constructing a fuzzy comparison matrix, synthesizing judgments and checking for consistency.

(a) **Constructing a Fuzzy Comparison Matrix**

The fuzzy matrix (A) of pair-wise comparisons is constructed from $i \times j$ elements, where i and j are the number of criteria (n). Let $A = \left[\tilde{a}_{ij}\right]_{n \times n}$ be a preference matrix such that $\tilde{a}_{ij} = \left(\tilde{a}_{ij}^l, \tilde{a}_{ij}^m, \tilde{a}_{ij}^n\right)$ is a triangular fuzzy number (TFN). In this matrix, a_{ij} represents the linguistic value of comparing of the i criterion with respect to the j criterion, as follows:

$$A = \begin{bmatrix} \tilde{a}_{12} & \tilde{a}_{12} & \cdots & \tilde{a}_{1j} \\ \tilde{a}_{21} & \tilde{a}_{22} & \cdots & \tilde{a}_{2j} \\ \vdots & \vdots & \cdots & \vdots \\ \tilde{a}_{i1} & \tilde{a}_{i2} & \cdots & \tilde{a}_{ij} \end{bmatrix} \tag{6.13}$$

The linguistic value can be determined using TFN. A TFN is denoted as (l, m, n), where l, m and n represent, respectively, the smallest value, the most promising value, and the largest possible value of comparing of the i criterion with respect to the j criterion. $\tilde{a}_{ij} = \tilde{a}_{ji}^{-1} = \left(l_{ij}, m_{ij}, n_{ij}\right)^{-1} = \left(\frac{1}{n_{ij}}, \frac{1}{m_{ij}}, \frac{1}{l_{ij}}\right)$, and $\tilde{a}_{ij} = (1, 1, 1)$ when $i = j$, as follows:

$$A = \begin{bmatrix} (1,1,1) & (l_{12}, m_{12}, n_{12}) & \cdots & (l_{1j}, m_{1j}, n_{1j}) \\ \left(\frac{1}{n_{21}}, \frac{1}{m_{21}}, \frac{1}{l_{21}}\right) & (1,1,1) & \cdots & (l_{2j}, m_{2j}, n_{2j}) \\ \vdots & \vdots & \cdots & \vdots \\ \left(\frac{1}{n_{i1}}, \frac{1}{m_{i1}}, \frac{1}{l_{i1}}\right) & \left(\frac{1}{n_{i2}}, \frac{1}{m_{i2}}, \frac{1}{l_{i2}}\right) & \cdots & (1,1,1) \end{bmatrix} \tag{6.14}$$

(b) *Synthesizing Judgments*

This stage represents a process for calculating the weight of each criterion. This process can be achieved as follows:

Step 1: *Calculation of fuzzy synthetic extent*
The value of fuzzy synthetic extent with respect to the ith is defined as follows:

$$S_i = \sum_{j=1}^{m} \tilde{a}_{ij} \times \left[\sum_{i=1}^{n} \sum_{j=1}^{m} \tilde{a}_{ij} \right]^{-1} \tag{6.15}$$

$\sum_{j=1}^{m} \tilde{a}_{ij}$, is obtained from fuzzy addition operation of m extent analysis value for a particular matrix such that

$$\sum_{j=1}^{m} \tilde{a}_{ij} = \sum_{j=1}^{m} l_j, \sum_{j=1}^{m} m_j, \sum_{j=1}^{m} n_j \tag{6.16}$$

and $\left[\sum_{i=1}^{n} \sum_{j=1}^{n} \tilde{a}_{ij} \right]^{-1}$ is obtained from fuzzy addition operation of $a_{ij}(j = 1, 2, \ldots, m)$ values such that

$$\sum_{i=1}^{n} \sum_{j=1}^{m} \tilde{a}_{ij} = \left(\sum_{i=1}^{n} l_{ij}, \sum_{i=1}^{n} m_{ij}, \sum_{i=1}^{n} n_{ij} \right) \tag{6.17}$$

Finally, the inverse of the above vector is computed as

$$\left[\sum_{i=1}^{n} \sum_{j=1}^{m} \tilde{a}_{ij} \right]^{-1} = \left(\frac{1}{\sum_{i=1}^{n} n_{ij}}, \frac{1}{\sum_{i=1}^{n} m_{ij}}, \frac{1}{\sum_{i=1}^{n} l_{ij}} \right) \tag{6.18}$$

Step 2: *Comparison of fuzzy values*
As $\tilde{a}_2$ and $\tilde{a}_1$ are two triangular fuzzy numbers, the degree of possibility of $\tilde{a}_2 = (l_2, m_2, n_2) \geq \tilde{a}_1 = (l_1, m_1, n_1)$ is defined as

$$V(a_2 \geq a_1) = \sup_{X \geq Y} \left[min \left(\mu_{a_1}(x), \mu_{a_2}(y) \right) \right] \tag{6.19}$$

This is equivalent to

$$V(a_2 \geq a_1) = hgt(a_1 \cap a_2) = \mu_{a_2}(d) \tag{6.20}$$

where d is the ordinate of the highest intersection point between μ_{a_1} and μ_{a_2} (Fig. 6.11). When $\tilde{a}_2 = (l_2, m_2, n_2)$ and $\tilde{a}_1 = (l_1, m_1, n_1)$, then $\mu_{a_2}(d)$ is defined as follows:

Fig. 6.11 Interaction
between a_1 and a_2

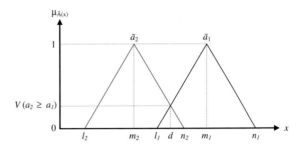

For the comparison between a_1 and a_2, the values of both $V(a_2 \geq a_1)$ and $V(a_1 \geq a_2)$ are required. The intersection between a_1 and a_2 is shown in Fig. 6.11.

Step 3: *Calculation of priority weights*
The degree of possibility for a convex fuzzy number to be greater than k convex fuzzy numbers $a_i (i = 1, 2, \ldots k)$ can be defined as

$$
\begin{aligned}
V(a \geq a_1, a_1, \ldots . a_k) &= V[(a \geq a_1) \, and \ldots and \, (a \geq a_k)] \\
&= \min V(a \geq a_i), \quad i = 1, 2, \ldots . k.
\end{aligned}
\tag{6.22}
$$

Assuming that

$$
d'(C_i) = \min V(S_i \geq S_k), \quad k = 1, 2, \ldots . n; \ k \neq i
\tag{6.23}
$$

then the weight vector is given by

$$
W' = \{ d'(C_1), d'(C_2), \ldots d'(C_n) \}^T
\tag{6.24}
$$

where $C_1, C_2, \ldots C_n$ are n criteria.

Step 4: *Calculation of normalized weight vectors*
The normalized weight vectors are obtained as follows, where W is a nonfuzzy number

$$
W = \{ d(C_1), d(C_2), \ldots d(C_n) \}^T
\tag{6.25}
$$

(c) ***Checking for consistency***

In order to identify the consistency of the linguistic values in the fuzzy matrix (A), the consistency ratio (CR) is calculated. If CR is less than or equal to 0.10, the linguistic values are acceptable. If CR is more than 0.10, the linguistic values are unacceptable and need to be altered. CR can be calculated as follows:

Table 6.2 The random consistency index

Size (n)	1	2	3	4	5	6	7	8
Random index (RI)	0	0	0.58	0.9	1.12	1.24	1.32	1.41

Step 1: *Compute the maximum eigenvalue (λ_{max})*
λ_{max} can be computed by calculating the consistency value (CV) of each row, then dividing the summation of CV by n to obtain λ_{max} as shown in Eq. 6.26.

$$\lambda_{max} = \frac{\sum_{i=1}^{n} CV_i}{n} \tag{6.26}$$

Step 2: *Calculate the consistency index (CI), using the following equation.*

$$CI = \frac{(\lambda_{max} - n)}{(n - 1)} \tag{6.27}$$

Step 3: *Calculate the consistency ratio (CR)*
CR can be computed by dividing the CI by the random index (RI). The value of RI depends on n. The RI values corresponding with n are listed in Table 6.2.

6.5.2 Methodology of FTOPSIS

FTOPSIS is the algorithm for finding the best option from a set of feasible alternatives. The main stages of FTOPSIS can be described as follows:

(a) **Establish fuzzy decision matrix**

The first step of the Fuzzy TOPSIS is to construct the $m \times n$ fuzzy-decision matrix ($\tilde{D}$).

$$\tilde{D} = \begin{array}{c} \\ A_1 \\ A_2 \\ \vdots \\ A_m \end{array} \begin{array}{cccc} C_1 & C_2 & \cdots & C_n \\ \left[\begin{array}{cccc} x_{11} & x_{12} & \cdots & x_{1n} \\ x_{21} & x_{22} & \cdots & x_{2n} \\ \vdots & \vdots & \ddots & \vdots \\ x_{m1} & x_{m2} & \cdots & x_{mn} \end{array}\right] \end{array} \tag{6.28}$$

where i is the criterion index ($i = 1, 2 \ldots m$), j is the alternative index ($j = 1, 2 \ldots n$). $C_1, C_2, \ldots, C_n$ denote the criteria; and $A_1, A_2, \ldots, A_m$ denote the possible alternatives. The elements $x_{ij} = (l_{ij}, m_{ij}, n_{ij})$ of the matrix are represented by the linguistic variables for the alternative j with respect to criterion i.

(b) *Normalize the fuzzy decision matrix*

The raw data are normalized using linear scale transformation to bring the different measurement units and scales into a comparable scale. The normalized fuzzy-decision matrix $\tilde{R}$ is calculated as:

$$\tilde{R} = \left[\tilde{r}_{ij}\right]_{m\times n}, \quad i = 1, 2\ldots m; \ j = 1, 2, \ldots n \tag{6.29}$$

Then the normalization process, for benefit related criteria (B) and for cost related criteria (C), can be performed using the Eqs. (6.30) and (6.31) respectively.

$$\tilde{r}_{ij} = \left(\frac{l_{ij}}{n_{ij}}, \frac{m_{ij}}{n_{ij}}, \frac{n_{ij}}{n_{ij}}\right), \ j\varepsilon B \tag{6.30}$$

$$\tilde{r}_{ij} = \left(\frac{l_{ij}}{l_{ij}}, \frac{m_{ij}}{l_{ij}}, \frac{n_{ij}}{l_{ij}}\right), \ j\varepsilon C \tag{6.31}$$

(c) *Compute the weighted decision matrix*

After calculating the weights of the evaluation criteria (w_j) using Fuzzy AHP, the weighted decision matrix ($\tilde{V}$) can be computed by multiplying the w_j with the values of the normalised fuzzy-decision matrix ($\tilde{r}_{ij}$). The $\tilde{V}$ is defined as

$$\tilde{V} = \left[\tilde{v}_{ij}\right]_{m\times n} = \left[\tilde{r}_{ij} \times \tilde{w}_{ij}\right]_{m\times n}, \quad i = 1, 2\ldots m; \ j = 1, 2, \ldots n \tag{6.32}$$

(d) *Determine the positive-ideal solution and negative-ideal solution*

The fuzzy positive ideal solution (FPIS) denoted by A^+ and the fuzzy negative ideal solution (FNIS) denoted by A^- are defined by the equations given below:
FPIS A^+ (aspiration levels)

$$A^+ = \left\{\tilde{v}_1^+, \ \tilde{v}_2^+, \ \ldots, \ \tilde{v}_j^+\right\} = \left\{(max_i v_{ij}|i = 1, 2, \ldots, m), \quad j = 1, 2, \ldots n\right\} \tag{6.33}$$

FNIS A^- (the worst levels)

$$A^- = \left\{\tilde{v}_1^-, \ \tilde{v}_2^-, \ \ldots, \ \tilde{v}_j^-\right\} = \left\{(min_i v_{ij}|i = 1, 2, \ldots, m), j = 1, 2, \ldots n\right\} \tag{6.34}$$

where $\tilde{v}_1^+ = (1, 1, 1)$ and $\tilde{v}_1^- = (0, 0, 0)$ for the benefit criteria, and $\tilde{v}_1^+ = (0, 0, 0)$ and $\tilde{v}_1^- = (1, 1, 1)$ for the cost criteria.

(e) *Calculate the distance of each alternative*

The distance (d_i^+, d_i^-) of each alternative from A^+ and A^- can be calculated as follows:

$$d_i^+ = \sum_{j=1}^{n} d\left(\tilde{v}_{ij}, \tilde{v}_j^+\right), \quad i = 1, 2, \ldots m \tag{6.35}$$

$$d_i^- = \sum_{j=1}^{n} d\left(\tilde{v}_{ij}, \tilde{v}_j^-\right), \quad i = 1, 2, \ldots m \tag{6.36}$$

where $d\left(\tilde{v}_{ij}, \tilde{v}_j^+\right)$ and $d\left(\tilde{v}_{ij}, v_j^-\right)$ denote the distance between two fuzzy numbers as given by the following equations:

$$d\left(\tilde{v}_{ij}, \tilde{v}_j^+\right) = \sqrt{\frac{1}{3}\left[(1 - l_1)^2 + (1 - m_1)^2 + (1 - n_1)^2\right]} \tag{6.37}$$

$$d\left(\tilde{v}_{ij}, \tilde{v}_j^-\right) = \sqrt{\frac{1}{3}\left[(0 - l_1)^2 + (0 - m_1)^2 + (0 - n_1)^2\right]} \tag{6.38}$$

(f) *Calculate the closeness coefficient*

The relative closeness coefficient (CC_j) of each alternative represents the distances to A^+ and A^- simultaneously. The CC_j can be calculated as follows:

$$CC_j = \frac{d_j^-}{d_j^+ + d_j^+}, \quad j = 1, \ldots, n \tag{6.39}$$

The different alternatives can be ranked in descending order according to the value of CC_j. The alternative with highest CC_j will be the best choice.

6.6   Concluding Remarks

This chapter showed that the integrating of multi-criteria decision-making (MCDM) methodology with fuzzy set theory (fuzzy MCDM) has the ability to: (1) handle any complex problem when the information is uncertain, and (2) reflect the human thinking style. Therefore, a hybrid fuzzy MCDM approach has been developed to achieve multi-objective optimization of dynamic scheduling in RFAC. The developed approach was divided into three phases: problem description; application of fuzzy methods; and analysis of the results.

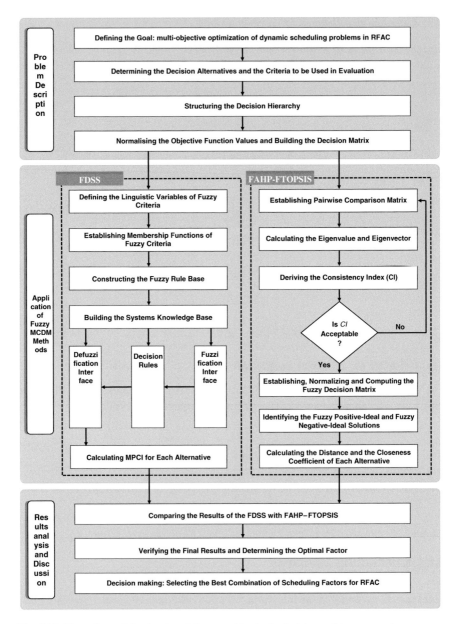

Fig. 6.12 Flow chart of the developed fuzzy multi criteria decision making approach

In the problem description phase, the steps for converting any MCDM problem to a hierarchical structure were described. This hierarchy is constructed based on three elements: the overall goal, criteria and sub-criteria (objective functions), and the decision alternatives (feasible solutions). The procedure for creating the fuzzy decision matrix is also described in this phase.

In the application of the fuzzy MCDM phase, the fuzzy decision support system (FDSS) and the fuzzy AHP-fuzzy TOPSIS (FAHP-FTOPSIS) are applied, to check the reliability and the consistency of the developed approach by comparing the obtained results of the proposed FDSS with FAHP-FTOPSIS.

- FDSS: this approach is built as a computer-based system using the Matlab fuzzy logic toolbox for obtaining the input data of the decision problem and for computing and visualizing the achieved results. The FDSS is designed to calculate the multiple performance characteristics index (MPCI), in order to find the optimal solution of multi-objective problems for dynamic scheduling in RFAC.
- FAHP-FTOPSIS: in this approach, the FAHP and FTOPSIS is integrated, as a new application, to optimize the dynamic scheduling in RFAC under different objective-functions. The FAHP is used to determine relative weights for multi scheduling criteria; and the FTOPSIS is proposed to evaluate the feasible solutions and make a final decision.

In the analysis phase, the results of FDSS are examined together with FAHP-FTOPSIS results, to check whether the FDSS results are reliable, or not. In addition, different analysis tools are used in this phase, to verify and validate the FDSS outcomes. These tools are: sensitivity analysis, analysis of mean, and percentage prediction error. A confirmation test is also performed to check whether any improved results can be obtained.

The following flow chart (Fig. 6.12) summarizes the three phases that are followed to develop the hybrid fuzzy MCDM approach. This approach will be later applied and validated using a realistic case study in Chap. 7.

References

Abd, K., Abhary, K., & Marian, R. (2011). An MCDM approach to selection scheduling rule in robotic flexible assembly cells. *World Academy of Science, Engineering and Technology, 76* (1), 643–648.
Abd, K., Abhary, K., & Marian, R. (2013a). Comparative analysis of MCDM methods and implementation of the scheduling rule selection problem: a case study in robotic flexible assembly cells. In *International Conference on Advanced Engineering Optimization Through Intelligent Techniques*, India, pp. 16–20.
Abd, K., Abhary, K., & Marian, R. (2013b). Fuzzy decision support system for selecting the optimal scheduling rule in robotic flexible assembly cells. *Australian Journal of Multi-Disciplinary Engineering, 9*(2), 125–132.
Abdullah, L. (2013). Fuzzy multi criteria decision making and its applications: A brief review of category. *Procedia—Social and Behavioral Sciences, 97*, 131–136.
Adibi, M. A., Zandieh, M., & Amiri, M. (2010). Multi-objective scheduling of dynamic job shop using variable neighborhood search. *Expert Systems with Applications, 37*(1), 282–287.

Aktan, H. E., & Tosun, Ö. (2013). An integrated fuzzy AHP—fuzzy TOPSIS approach for AS/RS selection. *International Journal of Productivity and Quality, 11*(2), 228–245.

Amiri, P. M. (2010). Project selection for oil-fields development by using the AHP and fuzzy TOPSIS methods. *Expert Systems with Applications, 37*(9), 6218–6224.

Awasthi, A., & Chauhan, S. S. (2012). A hybrid approach integrating affinity diagram, AHP and fuzzy TOPSIS for sustainable city logistics planning. *Applied Mathematical Modelling, 36*(2), 573–584.

Ayag, Z. (2005). A fuzzy AHP-based simulation approach to concept evaluation in a NPD environment. *IIE Transactions, 37*(9), 827–842.

Aydogan, E. K. (2011). Performance measurement model for Turkish aviation firms using the rough—AHP and TOPSIS methods under fuzzy environment. *Expert Systems with Applications, 38*(4), 3992–3998.

Azadegan, A., Porobic, L., Ghazinoory, S., Samouei, P., & Kheirkhah, A. S. (2011). Fuzzy logic in manufacturing: A review of literature and a specialized application. *International Journal of Production Economics, 132*(2), 258–270.

Azadeh, A., Nazari-Shirkouhi, S., Hatami-Shirkouhi, L., & Ansarinejad, A. (2011). A unique fuzzy multi-criteria decision making: computer simulation approach for productive operators' assignment in cellular manufacturing systems with uncertainty and vagueness. *International Journal of Advanced Manufacturing Technology, 56*(1), 329–343.

Berihaa, G. S., Patnaika, B., Mahapatraa, S. S., & Padheeb, S. (2012). Assessment of safety performance in Indian industries using fuzzy approach. *Expert Systems with Applications, 39*(3), 3311–3323.

Beskese, A., Kahraman, C., & Irani, Z. (2004). Quantification of flexibility in advanced manufacturing systems using fuzzy concept. *International Journal of Production Economics, 89*(1), 45–56.

Boran, F., Genc, S., Kurt, M., & Akay, D. (2009). A multi-criteria intuitionistic fuzzy group decision making for supplier selection with TOPSIS method. *Expert Systems with Applications, 36*(8), 11363–11368.

Bozdağ, C. E., Kahraman, C., & Ruan, D. (2003). Fuzzy group decision making for selection among computer integrated manufacturing systems. *Computers in Industry, 51*(1), 13–29.

Brugha, C. M. (2004). Structure of multi-criteria decision-making. *Journal of the Operational Research Society. 55*(11), 1156–1168.

Bugarski, V., Backalic, T., & Kuzmanov, U. (2013). Fuzzy decision support system for ship lock control. *Expert Systems with Applications, 40*(10), 3953–3960.

Buyukozkan, G., & Cifci, G. (2012). A combined fuzzy AHP and fuzzy TOPSIS based strategic analysis of electronic service quality in healthcare industry. *Expert Systems with Applications, 39*(13), 2341–2354.

Buyukozkan, G., Kahraman, C., & Ruan, D. (2004). A fuzzy multi-criteria decision approach for software development strategy selection. *International Journal of General Systems, 33*(2), 259–280.

Camastra, F., Ciaramella, A., Giovannelli, V., Lener, M., Rastelli, V., Staiano, A., et al. (2014). A fuzzy decision support system for the environmental risk assessment of genetically modified organisms. *Smart Innovation, Systems and Technologies, 26*, 241–249.

Chamodrakas, I., Batis, D., & Martakos, D. (2010). Supplier selection in electronic marketplaces using satisficing and fuzzy AHP. *Expert Systems with Applications. 37*(1), 490–498.

Chan, F. T. S., Kumar, N., Tiwari, M. K., Lau, H. C. W., & Choy, K. L. (2008). Global supplier selection: a fuzzy-AHP approach. *International Journal of Production Research, 46*(14), 3825–3857.

Choudhary, D., & Shankar, R. (2012). An STEEP-fuzzy AHP-TOPSIS framework for evaluation and selection of thermal power plant location: A case study from India. *Energy, 42*(1), 510–521.

Chu, T. C., & Lin, Y. C. (2003). A fuzzy TOPSIS method for robot selection. *International Journal of Manufacturing Technology, 21*(4), 284–290.

Dagdeviren, M. (2008). Decision making in equipment selection: an integrated approach with AHP and PROMETHEE. *Journal of Intelligent Manufacturing, 19*(4), 397–406.

Dağdeviren, M., Yavuz, S., & Kılınç, N. (2009). Weapon selection using the AHP and TOPSIS methods under fuzzy environment. *Expert Systems with Applications, 36*(4), 8143–8151.

Damghani, K. K., Sadi-Nezhad, S., & Aryanezhad, M. B. (2011). A modular Decision Support System for optimum investment selection in presence of uncertainty: Combination of fuzzy mathematical programming and fuzzy rule based system. *Expert Systems with Applications, 38*(1), 824–834.

Deb, S. K., & Bhattacharyya, B. (2005). Fuzzy decision support system for manufacturing facilities layout planning. *Decision Support Systems, 40*(2), 305–314.

Duran, O., & Aguilo, J. (2008). Computer-aided machine-tool selection based on a Fuzzy-AHP approach. *Expert Systems with Applications, 34*(3), 1787–1794.

Ekmekçioğlu, M., Kaya, T., & Kahraman, C. (2010). Fuzzy multicriteria disposal method and site selection for municipal solid waste. *Waste Management, 30*(8), 1729–1736.

Fasanghari, M., & Montazer, G. A. (2010). Design and implementation of fuzzy expert system for Tehran Stock Exchange portfolio recommendation. *Expert Systems with Applications, 37*(9), 6138–6147.

Fattahi, P., & Fallahi, A. (2010). Dynamic scheduling in flexible job shop systems by considering simultaneously efficiency and stability. *CIRP Journal of Manufacturing Science and Technology, 2*(2), 114–123.

Figueira, J., Greco, S., & Ehrgott, M. (2005). *Multiple criteria decision analysis: State of the art surveys*. New York, USA: Springer.

García, N., Puente, J., Fernández, I., & Priore, P. (2013). Supplier selection model for commodities procurement. Optimised assessment using a fuzzy decision support system. *Applied Soft Computing, 13*(4), 1939–1951.

Guarnieri, P. (2014). Decision making regarding information sharing in collaborative relationships under an MCDA perspective. *International Journal of Management and Decision Making, 13*(1), 77–98.

Gumus, A. T. (2009). Evaluation of hazardous waste transportation firms by using a two step fuzzy-AHP and TOPSIS methodology. *Expert Systems with Applications, 36*(2), 4067–4074.

Haji, A., & Assadi, M. (2009). 'Fuzzy expert systems and challenge of new product pricing. *Computers and Industrial Engineering, 56*(2), 616–630.

Hesami, M., Nosratabadi, H. E., & Fazlollahtabar, H. (2012). Design of a fuzzy expert system for determining adjusted price of products and services. *International Journal of Industrial and Systems Engineering, 13*(1), 1–26.

Hwang, C., & Yoon, K. (1981). *Multiple attribute decision making methods and application*. New York: Springer.

Hyde, K. M. (2006). *Uncertainty analysis methods for multi-criteria decision analysis*. Doctor of Philosophy: University of Adelaide.

Jafarian, M., & Vahdat, S. E. (2012). A fuzzy multi-attribute approach to select the welding process at high pressure vessel manufacturing. *Journal of Manufacturing Processes, 14*(3), 250–256.

Javanbarg, M. B., Scawthorn, C., Kiyono, J., & Shahbodaghkhan, B. (2012). Fuzzy AHP-based multicriteria decision making systems using particle swarm optimization. *Expert Systems with Applications, 39*(1), 960–966.

Kaboli, A., Aryanezhad, M., Shahanaghi, K., & Niroomand, I. (2007), A new method for plant location selection problem: A fuzzy AHP approach. In *Proceedings of the IEEE International Conference on System, Man and Cybernetics*, pp. 582–586.

Kahraman, C., Cebeci, U., & Ruan, D. (2004). Multi-attribute comparison of catering service companies using fuzzy AHP: The case of Turkey. *International Journal of Production Economics, 87*(2), 171–184.

Kahraman, C., Beskese, A., & Kaya, I. (2010). Selection among ERP outsourcing alternatives using a fuzzy multi-criteria decision making methodology. *International Journal of Production Research, 48*(2), 547–566.

Kang, H. Y., & Lee, A. H. I. (2007). Priority mix planning for semiconductor fabrication by fuzzy AHP ranking. *Expert Systems with Applications, 32*(2), 560–570.

Korhonen, P., Moskowitz, H., & Wallenius, J. (1992). Multiple criteria decision support—A review. *European Journal of Operational Research, 63*(3), 361–375.

Kumar, P., & Singh, R. K. (2012). A fuzzy AHP and TOPSIS methodology to evaluate 3PL in a supply chain. *Journal of Modelling in Management, 7*(3), 287–303.

Kutlu, A. C., & Ekmekcioglu, M. (2012). Fuzzy failure modes and effects analysis by using fuzzy TOPSIS-based fuzzy AHP. *Expert Systems with Applications, 39*(1), 61–67.

Lee, K. (2005). *First course on fuzzy theory and applications*. Berlin: Springer-Verlag.

Lin, H. F. (2010). An application of fuzzy AHP for evaluating course website quality. *Computers and Education, 54*(4), 877–888.

Lin, M.C., Wang, C.C., Chen, M.S. & Chang, C.A. (2008). Using AHP and TOPSIS approaches in customer-driven product design process. *Computers in Industry, 59*(1), 17–31.

Maldonado-Macías, A., Alvarado, A., García, J. L., & Balderrama, C. O. (2014). Intuitionistic fuzzy TOPSIS for ergonomic compatibility evaluation of advanced manufacturing technology. *International Journal of Advanced Manufacturing Technology, 70*(9–12), 2283–2292.

Mentes, A., & Helvacioglu, I. H. (2012). Fuzzy decision support system for spread mooring system selection. *Expert Systems with Applications, 39*(3), 3283–3297.

Montgomery, D. C. (2004). *Design and analysis of experiments* (5th ed.). New York: Wiley.

Önüt, S., & Soner, S. (2008). Transshipment site selection using the AHP and TOPSIS approaches under fuzzy environment. *Waste Management, 28*(9), 1552–1559.

Paksoy, T., Pehlivan, N. Y., & Kahraman, C. (2012). Organizational strategy development in distribution channel management using fuzzy AHP and hierarchical fuzzy TOPSIS. *Expert Systems with Applications, 39*(3), 2822–2841.

Pandey, A. K., & Dubey, A. K. (2012). Taguchi based fuzzy logic optimization of multiple quality characteristics in laser cutting of Duralumin sheet. *Optics and Lasers in Engineering, 50*, 328–335.

Parsaei, S., Keramati, M. A., Zorriassatine, F., & Feylizadeh, M. R. (2012). An order acceptance using FAHP and TOPSIS methods: A case study of Iranian vehicle belt production industry selection. *International Journal of Industrial Engineering Computations, 3*(2), 211–224.

Patil, S. K., & Kant, R. (2014). A fuzzy AHP-TOPSIS framework for ranking the solutions of knowledge management adoption in supply chain to overcome its barriers. *Expert Systems with Applications, 41*(2), 679–693.

Phadke, M. S. (1989). *Quality engineering using robust design*. Englewood Cliffs, NJ, USA: Prentice-Hall.

Rangsaritratsamee, R., Ferrell, W. G, Jr, & Kurz, M. B. (2004). Dynamic rescheduling that simultaneously considers efficiency and stability. *Computers and Industrial Engineering, 46*(1), 1–15.

Rao, R. V. (2011). *Advanced modeling and optimization of manufacturing processes: International research and development*. London: Springer.

Rostamzadeh, R., & Sofian, S. (2011). Prioritizing effective 7 Ms to improve production systems performance using fuzzy AHP and fuzzy TOPSIS (case study). *Expert Systems with Applications, 38*(5), 5166–5177.

Saaty, T. L. (1980). *The analytic hierarchy process, planning, priority setting, resource allocation*. New York: McGraw-Hill.

Samvedi, A., Jain, V., & Chan, F. T. S. (2013). Quantifying risks in a supply chain through integration of fuzzy AHP and fuzzy TOPSIS. *International Journal of Production Research, 51*(8), 2433–2442.

Seçme, N. Y., Bayrakdaroglu, A., & Kahraman, C. (2009). Fuzzy performance evaluation in Turkish banking sector using analytic hierarchy process and TOPSIS. *Expert Systems with Applications, 36*(9), 11699–11709.

Shyjith, K., Ilangkumaran, M., & Kumanan, S. (2008). Multi-criteria decision-making approach to evaluate optimum maintenance strategy in textile industry. *Journal of Quality in Maintenance Engineering, 14*(4), 375–386.

Sun, C. C. (2010). A performance evaluation model by integrating fuzzy AHP and fuzzy TOPSIS methods. *Expert Systems with Applications, 37*(12), 7745–7754.

Tavakkoli-Moghaddam, R., Azarkish, M., & Sadeghnejad-Barkousaraie, A. (2011a). A new hybrid multi-objective Pareto archive PSO algorithm for a bi-objective job shop scheduling problem. *Expert Systems with Applications, 38*(9), 10812–10821.

Tavakkoli-Moghaddam, R., Azarkish, M., & Sadeghnejad-Barkousaraie, A. (2011b). Solving a multi-objective job shop scheduling problem with sequence-dependent setup times by a Pareto archive PSO combined with genetic operators and VNS. *The International Journal of Advanced Manufacturing Technology, 53*(5–8), 733–750.

Tolga, E., Demircan, M. L., & Kahraman, C. (2005). Operating system selection using fuzzy replacement analysis and analytic hierarchy process. *International Journal of Production Economics, 97*(1), 89–117.

Torfi, F., Farahani, R. Z., & Rezapour, S. (2010). Fuzzy AHP to determine the relative weights of evaluation criteria and Fuzzy TOPSIS to rank the alternatives. *Applied Soft Computing, 10*(2), 520–528.

Triantaphyllou, E. (2000). *Multi-criteria decision making methods: a comparative study.* Dordrecht: Kluwer Academic Publishers.

Vaidya, O. S., & Kumar, S. (2006). Analytic hierarchy process: An overview of applications. *European Journal of Operational Research, 169*(1), 1–29.

Wang, L. X. (1997). *A course in fuzzy systems and control.* Upper Saddle River, NJ: Prentice-Hall.

Wang, Y. M., Luo, Y., & Hua, Z. (2008). On the extent analysis method for fuzzy AHP and its applications. *European Journal of Operational Research, 186*(12), 735–747.

Wang, J., Fan, K., & Wang, W. (2010). Integration of fuzzy AHP and FPP with TOPSIS methodology for aeroengine health assessment. *Expert Systems with Applications, 37*(12), 8516–8526.

Xing, L. N., Chen, Y. W., & Yang, K. W. (2009). An efficient search method for multi-objective flexible job shop scheduling problems. *Journal of Intelligent Manufacturing, 20*(3), 283–293.

Yong, D. (2006). Plant location selection based on fuzzy TOPSIS. *International Journal of Advanced Manufacturing Technology, 28*(7–8), 839–844.

Yu, V. F., & Hu, K. J. (2010). An integrated fuzzy multi-criteria approach for the performance evaluation of multiple manufacturing plants. *Computers and Industrial Engineering, 58*(2), 269–277.

Yu, X., Guo, S., Guo, J. & Huang, X. (2011). Rank B2C e-commerce websites in e-alliance based on AHP and fuzzy TOPSIS. *Expert Systems with Applications*, 38 (4), 3550–3557.

Zhang, L., Li, X., Gao, L., & Zhang, G. (2013). Dynamic rescheduling in FMS that is simultaneously considering energy consumption and schedule efficiency. *International Journal of Advanced Manufacturing Technology,*. doi:10.1007/s00170-013-4867-3.

Zouggari, A., & Benyoucef, L. (2012). Simulation based fuzzy TOPSIS approach for group multi-criteria supplier selection problem. *Engineering Applications of Artificial Intelligence, 25*(3), 507–519.

Chapter 7
Case Study 2: Application of Hybrid Fuzzy MCDM Approach to Optimize Dynamic Scheduling in RFAC

7.1 Introduction

The research work described in Chap. 5 was to propose a methodology for scheduling RFAC in a dynamic environment by combining Taguchi experimental design method with simulation modelling. This work was restricted to solving single-objective optimization problems. Chapter 6 was devoted to achieving multi-objective optimisation of dynamic scheduling in RFAC using a new Fuzzy Multi-Criteria Decision-Making (Fuzzy MCDM) approach. The proposed approach will be examined, verified and validated in this chapter, using a realistic case study.

The objective of this chapter is to demonstrate the application of the proposed approach to a comprehensive decision problem for multi-objective optimization of dynamic scheduling problems in RFAC. In order to achieve the stated objective, the following five steps have to be performed sequentially:

- Implement the proposed hybrid fuzzy approach using the fuzzy decision support system (FDSS) and fuzzy AHP-fuzzy TOPSIS (FAHP-FTOPSIS).
- Perform a statistical analysis to demonstrate the stability and robustness of FDSS and FAHP-FTOPSIS.
- Check the reliability and consistency of the proposed approach by comparing the obtained results of the proposed FDSS with FAHP-FTOPSIS.
- Predict the optimal factor combinations that optimize the objective functions, and identify the most significant factors that affect the system's performance.
- Perform a sensitivity analysis and confirmation test to verify and validate the results obtained.

© Springer International Publishing Switzerland 2016
K.K. Abd, *Intelligent Scheduling of Robotic Flexible Assembly Cells*,
Springer Theses, DOI 10.1007/978-3-319-26296-3_7

7.2   Case Study

The objective of this case study is to examine the robustness of the hybrid approach presented in Chap. 6. This case study is a complement to the work presented in Chap. 5 (Sects. 5.4–5.6). Thus, the input data of this continuing case study represents the output (simulation results) in Chap. 5. In the simulation results, three independent objective functions, makespan (C_{max}), total tardiness (TD) and number of tardy jobs (N_T) were calculated. Tables 5.9, 5.10 and 5.11 showed the three objective functions with their signal-to-noise (S/N) ratios.

From these results, for single-objective scheduling, it can be seen that the optimum combination of scheduling factors is the one having the smallest objective function value (the highest S/N ratio). Nevertheless, for multi-objective scheduling, it can be concluded that the optimum combination of scheduling factors is not as straightforward as that of single-objective scheduling, because a higher S/N ratio for one objective response may correspond to a lower S/N ratio for another objective response. These simulation results reveal that the optimum combination of scheduling factors is a multi-criteria decision-making (MCDM) problem. Consequently, a hybrid fuzzy MCDM approach was developed in Chap. 6 to deal with this problem.

To simplify the complexity of the MCDM problem, it can be described as a hierarchical structure, as shown in Fig. 7.1. This hierarchy contains three main elements: *overall goal*, *criteria* and *alternatives*.

Overall goal: the main goal is to solve the multi-criteria decision-making problem for optimization of dynamic scheduling problems in RFAC.

Criteria: The criteria represent the objective functions to fulfil the overall goal. Three objective functions are considered in this problem. These objectives are classified into two types: time-based and due date-based objectives. The makespan (C_{max}), is in the first category while total tardiness (TD) and the number of tardy jobs (N_T) fall into the second category.

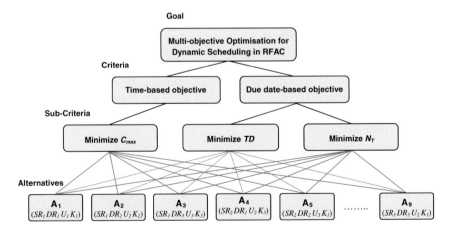

Fig. 7.1 The decision hierarchy

Table 7.1 Simulation results with their normalizations

Alternatives	C_{max}	S/N	μ_{Cmax}	TD	S/N	μ_{TD}	N_{TD}	S/N	μ_{NTD}
A_1 (SR_1 DR_1 U_1 K_1)	38,499	−91.71	0.32	10,070	−80.061	0.35	4	−12.041	0.71
A_2 (SR_1 DR_2 U_2 K_2)	34,736	−90.82	0.67	19,124	−85.632	0.21	12	−21.584	0.25
A_3 (SR_1 DR_3 U_3 K_3)	31,571	−89.99	1.00	50,161	−94.007	0.00	22	−26.848	0.00
A_4 (SR_2 DR_1 U_2 K_3)	35,387	−90.98	0.61	28,904	−89.219	0.12	18	−25.105	0.08
A_5 (SR_2 DR_2 U_3 K_1)	32,488	−90.23	0.90	5666.0	−75.066	0.48	2	−6.021	1.00
A_6 (SR_2 DR_3 U_1 K_2)	42,356	−92.54	0.00	9909.0	−79.921	0.36	7	−16.902	0.48
A_7 (SR_3 DR_1 U_3 K_2)	33,003	−90.37	0.85	2994.0	−69.525	0.62	6	−15.563	0.54
A_8 (SR_3 DR_2 U_1 K_3)	39,999	−92.04	0.19	31,792	−90.046	0.10	19	−25.575	0.06
A_9 (SR_3 DR_3 U_2 K_1)	36,310	−91.20	0.52	542.00	−54.680	1.00	2	−6.021	1.00

Alternatives: The decision alternatives represent the simulation experiments (feasible solutions) with respect to the multi-objective functions. These experiments are conducted under different combinations of scheduling factors: sequencing rule (*SR*), dispatching rule (*DR*), cell utilization (U) and due date tightness (K),

After the MCDM problem is constructed as a hierarchical structure, a decision table of the alternatives with respect to the conflicting objectives is built. In the decision table, the S/N ratios for the three objective functions, makespan (C_{max}) total tardiness (*TD*) and number of tardy jobs (N_T) must be within the same range. In this case study, the value ranges of the three objectives are all different, as shown in Tables 5.9, 5.10 and 5.11. To avoid the different ranges, the values of the S/N ratio must be normalized. Since the three objectives are the '*smaller the better*' type, the normalization is determined using Eq. (6.1). The normalization values are between 0 and 1; 0 means the least desirable value while 1 denotes the best value. The simulation results of the three objective functions with their normalizations are shown in Table 7.1.

7.3   Application Using FDSS

In order to obtain the optimum scheduling for multi-objective problems in RFAC, a FDSS is developed and applied. The FDSS combines multi-objective functions in one performance measure named multiple performance characteristics index (*MPCI*). The FDSS is built using the fuzzy logic toolbox in MATLAB software.

In this system, four steps (described in Sect. 6.4.2) are performed. The next sub-section will explain how these steps are implemented.

7.3.1   Defining Input and Output Variables

The FDSS contains three input variables and one output variable, according to the example application described in this chapter. The input variables are the S/N ratio of C_{max}, TD and N_T, while $MPCI$ is the output variable, representing the performance measure to evaluate the scheduling results of different experiments. The experiment with the highest $MPCI$ will be chosen as the optimal solution. Figure 7.2 shows the inference system of FDSS.

7.3.2   Specifying Input and Output Membership Functions

In this step, the input and output variables are constructed from the two most well-known of membership functions shapes (triangular and trapezoidal). The three input variables (C_{max}, TD and N_T) are built from triangular membership functions, as shown in Fig. 7.3. The values of the membership functions are between 0 and 1. These values are measured and assessed based on the S/N ratio results of the C_{max}, TD and N_T individually. The $MPCI$, which represents the fuzzy output of the FDSS, takes both triangular and trapezoidal membership functions, as shown in Fig. 7.4.

The S/N ratios of C_{max}, TD and N_T are broken into a set of linguistic values for the inputs: poor (P), satisfactory (S), and good (G); while the output is set into seven linguistic values: tiny (T), very small (VS), small (S), medium (M), large (L), very large (VL) and huge (H). Table 7.2 shows the different linguistic values of inputs/output and their numerical range.

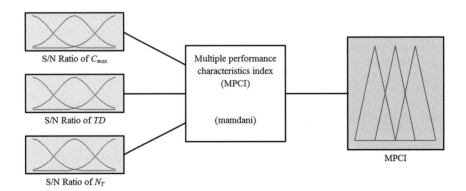

Fig. 7.2 The inference system of FDSS

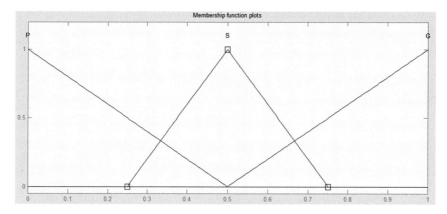

Fig. 7.3 Membership functions of linguistic input variables

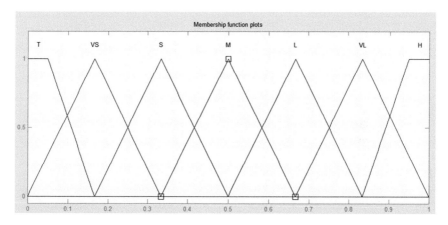

Fig. 7.4 Membership functions of linguistic output variables (MPCI)

7.3.3 Constructing Decision Rules and Knowledge Base

The proposed system makes decisions and generates output values based on knowledge provided by the decision makers in the form of IF-THEN fuzzy rules. The Fuzzy decision rules are designed to control the output variable (*MPCI*). Since the S/N ratios of C_{max}, *TD* and N_T have three states each, the total number of fuzzy decision rules is twenty seven ($3^3 = 27$), based on the formula (n^m), where n and m denote input variables (S/N Ratios) and their linguistic values respectively. Table 7.3 demonstrates the 27 fuzzy rules. Each rule is mathematically evaluated through a process named implication, and the results of all of the evaluations are analyzed and combined using fuzzy aggregations.

Table 7.2 Input and output variables with their fuzzy values

System variable	Linguistic variables	Linguistic value	Term set	Numerical range
Inputs	S/N ratios of C_{max}, TD and N_T	Poor	P	[0.00–0.50]
		Satisfactory	S	[0.25–0.75]
		Good	G	[0.50–1.00]
Output	MPCI	Tiny	T	[0.00–0.15]
		Very small	VS	[0.00–0.33]
		Small	S	[0.16–0.50]
		Medium	M	[0.33–0.66]
		Large	L	[0.50–0.83]
		Very large	VL	[0.66–1.00]
		Huge	H	[0.85–1.00]

Table 7.3 Fuzzy rules

Rule no.	S/N of C_{max}	S/N of TD	S/N of N_{TD}	MPCI
1	Poor	Poor	Poor	Tiny
2	Poor	Poor	Satisfactory	Very small
3	Poor	Poor	Good	Very small
4	Poor	Satisfactory	Poor	Very small
5	Poor	Satisfactory	Satisfactory	Small
6	Poor	Satisfactory	Good	Small
7	Poor	Good	Poor	Small
8	Poor	Good	Satisfactory	Small
9	Poor	Good	Good	Medium
10	Satisfactory	Poor	Poor	Small
11	Satisfactory	Poor	Satisfactory	Small
12	Satisfactory	Poor	Good	Medium
13	Satisfactory	Satisfactory	Poor	Medium
14	Satisfactory	Satisfactory	Satisfactory	Medium
15	Satisfactory	Satisfactory	Good	Medium
16	Satisfactory	Good	Poor	Medium
17	Satisfactory	Good	Satisfactory	Large
18	Satisfactory	Good	Good	Large
19	Good	Poor	Poor	Medium
20	Good	Poor	Satisfactory	Large
21	Good	Poor	Good	Large
22	Good	Satisfactory	Poor	Large
23	Good	Satisfactory	Satisfactory	Large
24	Good	Satisfactory	Good	Very large
25	Good	Good	Poor	Very large
26	Good	Good	Satisfactory	Very large
27	Good	Good	Good	Huge

7.3.4 Determining MPCI by Using Defuzzification

The last step in FDSS is to determine the *MPCI* of the evaluated experiments. Determination of *MPCI* can be performed via the fuzzified surface viewer and rule viewer. The surface viewer allows the user to visualize the relation between input fuzzy variables and the output of a fuzzy system (*MPCI*) in a three-dimensional graph. In this example, since the number of input variables is three, the number of generated 3D graphs is also three, as shown in Figs. 7.5, 7.6 and 7.7. These Figures show the effects of the combination of C_{max}, *TD and* N_T the *MPCI*.

Figure 7.5 depicts the priority response surface for makespan (C_{max}) and total tardiness (*TD*). It can be seen that the C_{max} has a higher influence than the *TD* on the *MPCI*. It also can be seen that the good C_{max} and *TD* give a high score for *MPCI*.

Figure 7.6 illustrates the *MPCI* from the perspective of makespan (C_{max}) and number of tardy jobs (N_T). In this figure, the makespan has a greater influence than the number of tardy jobs on the *MPCI*. It is also shown that the *MPCI* decreases with decreasing makespan and number performance.

Fig. 7.5 3D surface plots of the MPCI for makespan and total tardiness

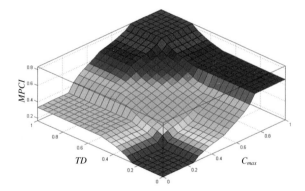

Fig. 7.6 3D surface plots of the MPCI for makespan and number of tardy jobs

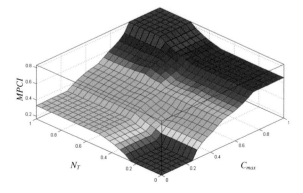

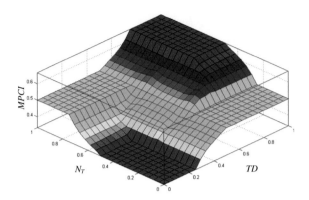

Fig. 7.7 3D surface plots of the MPCI for total tardiness and number of tardy jobs

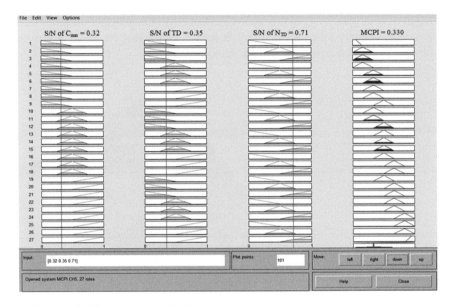

Fig. 7.8 Fuzzified rule viewer of the inputs/output

Figure 7.7 represents the *MPCI* from the perspective of total tardiness (*TD*) and number of tardy jobs (N_T). Again a good total tardiness performance, regardless of number of tardy jobs, gives a high *MPCI* score. Also, it can be seen that the number of tardy jobs has a smaller impact on *MPCI* value as compared to the total tardiness.

The fuzzified rule viewer, which displays a graphical representation of the C_{max}, *TD* and N_T through all the fuzzy rules, is shown in Fig. 7.8. This viewer can accept any value of the S/N ratio. The output (*MPCI*) in the fuzzified rule viewer can be interpreted easily, as in the following: IF the S/N of C_{max} is (0.32), the S/N of TD is (0.35) and S/N of N_{TD} is (0.71) THEN *MCPI* will be (0.330). Table 7.4 and Fig. 7.9 show the *MPCI* values corresponding to each experimental run.

Table 7.4 MPCI values obtained by FDSS

Alternatives	μ Cmax	μ TD	μ NT	MCPI
A_1 $(SR_1\ DR_1\ U_1\ K_1)$	0.32	0.35	0.71	0.330
A_2 $(SR_1\ DR_2\ U_2\ K_2)$	0.67	0.21	0.25	0.419
A_3 $(SR_1\ DR_3\ U_3\ K_3)$	1.00	0.00	0.00	0.500
A_4 $(SR_2\ DR_1\ U_2\ K_3)$	0.61	0.12	0.08	0.383
A_5 $(SR_2\ DR_2\ U_3\ K_1)$	0.90	0.48	1.00	0.824
A_6 $(SR_2\ DR_3\ U_1\ K_2)$	0.00	0.36	0.48	0.267
A_7 $(SR_3\ DR_1\ U_3\ K_2)$	0.85	0.62	0.54	0.723
A_8 $(SR_3\ DR_2\ U_1\ K_3)$	0.19	0.10	0.06	0.058
A_9 $(SR_3\ DR_3\ U_2\ K_1)$	0.52	1.00	1.00	0.676

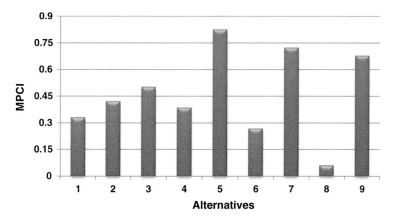

Fig. 7.9 Final ranking of alternatives using FDSS

7.4   Application Using FAHP-FTOPSIS

In this section, the FAHP is applied to determine the weights of the multiple evaluation criteria to be used in the evaluation process, and the FTOPSIS is applied for evaluation of each alternative scheduling combination based on their overall performance in order to make a final decision.

7.4.1   Application of FAHP in Determining Weights of Criteria

In Fuzzy Analytic Hierarchy Process (FAHP), the fuzzy comparison matrix is created by experts who have experience relevant to the related decision area. The experts decide how much one criterion dominates another. Hence, the experts are given the task of forming an individual pairwise comparison matrix by using triangular fuzzy number (TFN). In this research, TFN and the reciprocal scales are defined with the corresponding membership functions as shown in Fig. 7.10 and Table 7.5 respectively. TFNs are used to overcome shortcomings of the nine-point scale in the traditional AHP. Table 7.6 presents the aggregate fuzzy comparison matrix with respect to the overall objective.

After the fuzzy comparison matrix is constructed, the weight of each criterion can be calculated as follows: Firstly, the values of the fuzzy synthetic extent of the three criteria are calculated as below by using Eq. (6.15):

$$S_1 = (3.00, 7.00, 11.00) \times (19, 11.666, 6.733)^{-1} = (0.158, 0.600, 1.634)$$

$$S_2 = (2.20, 2.33, 5.00) \times (19, 11.666, 6.733)^{-1} = (0.116, 0.200, 0.743)$$

$$S_3 = (1.53, 2.33, 3.00) \times (19, 11.666, 6.733)^{-1} = (0.081, 0.200, 0.446)$$

Fig. 7.10 Linguistic variables for the importance weight of each criterion

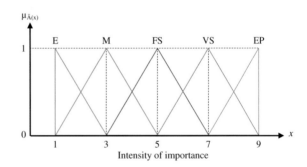

Table 7.5 Fuzzy evaluation scale for the weights

Linguistic scale for importance	Linguistic term	Triangular fuzzy scale	Triangular fuzzy reciprocal scale
Just equal		(1, 1, 1)	(1, 1, 1)
Equal importance	E	(1, 1, 3)	(1/3, 1, 1)
Moderate	M	(1, 3, 5)	(1/5, 1/3, 1)
Fairly strong	FS	(3, 5, 7)	(1/7, 1/5, 1/3)
Very strong	VS	(5, 7, 9)	(1/9, 1/7, 1/5)
Extremely preferred	EP	(7, 9, 9)	(1/9, 1/9, 1/9)

Table 7.6 The fuzzy comparison matrix of criteria

	Makespan (C_1)	Total tardiness (C_2)	Number of tardy jobs (C_3)
C_1	(1, 1, 1)	(1, 3, 5)	(1, 3, 5)
C_2	(1/5, 1/3, 1)	(1, 1, 1)	(1, 1, 3)
C_3	(1/5, 1/3, 1)	(1/3, 1, 1)	(1, 1, 1)

Secondly, the degrees of possibility of S_i over S_j ($i \neq j$) are calculated by Eq. (6.19).

$$V(S_1 \geq S_2) = 1.00,$$
$$V(S_1 \geq S_3) = 1.00,$$
$$V(S_2 \geq S_1) = \frac{(0.158 - 0.743)}{(0.200 - 0.743) - (0.600 - 0.158)} = 0.59,$$
$$V(S_2 \geq S_3) = 1.00,$$
$$V(S_3 \geq S_1) = \frac{(0.158 - 0.446)}{(0.200 - 0.446) - (0.600 - 0.158)} = 0.42,$$
$$V(S_3 \geq S_2) = 1.00$$

Thirdly, the minimum degree of possibility is determined using Eq. (6.22) as follows:

$$m(C_1) = \min V(S_i \geq S_k) = \min (1.00, 1.00) = 1.00$$
$$m(C_2) = \min V(S_i \geq S_k) = \min (0.59, 1.00) = 0.59$$
$$m(C_3) = \min V(S_i \geq S_k) = \min (0.42, 1.00) = 0.42$$

Therefore, the weight vector is given by: $Wc = (1.00, 0.59, 0.42)^T$.

Finally, after normalizing Wc, the normalized weight vectors of each of the three criteria can be determined as $W = (0.497, 0.295, 0.208)^T$. The makespan ($C_1$) is determined as the most important criterion in the combinations of scheduling factors, followed by total tardiness (C_2); the number of tardy jobs (C_3) appears to be less significant. After the weights of each criterion are calculated, the last stage in FAHP is to identify the consistency of the linguistic values in the fuzzy comparison matrix. In this research, the consistency ratio (CR) is calculated using Eqs. 6.26 and 6.27. The results obtained from the calculations based on the fuzzy comparison matrix provided in Table 7.6 are presented in Table 7.7. As can be seen from Table 7.7, the consistency ratio (CR) is less than 0.10. Thus, the weights are consistent and they are able to be used in the evaluation process.

Table 7.7 Results obtained from FAHP approach

Criteria	Weights (w)	λ_{max}	CI	RI	CR
C_1	0.497	3.0921	0.0145	0.58	0.0251
C_2	0.295				
C_3	0.208				

7.4.2   Application of FTOPSIS in Ranking of Alternatives

In FTOPSIS, six stages are required for evaluating and ranking the alternatives.

Stage 1: Establish the fuzzy decision matrix by evaluating alternatives under each of the criteria using linguistic variables. The membership functions of these linguistic variables are shown in Fig. 7.11. In this research, nine alternatives (scheduling factor combinations) must be evaluated with respect to each criterion, as shown in Table 7.1. The fuzzy decision matrix of the elements in Table 7.1 is demonstrated in Table 7.8.

Stage 2: Normalize the fuzzy decision matrix. Since the ranges of all the values in Table 7.8 belong to the closed interval [0, 1], there is no need for normalization.

Stage 3: Obtain a weighted decision matrix. This matrix can be established using the Eq. (6.32) and the criteria weights calculated by fuzzy AHP (Table 7.7). The weighted decision matrix is presented in Table 7.9.

Stage 4: Determine the fuzzy positive ideal solution (*FPIS*, A^+) and the fuzzy negative ideal solution (*NPIS*, A^-) using Eqs. 6.33 and 6.34. In this problem, C_1, C_2, C_3 are considered as benefit criteria, so $\tilde{v}_1^+ = (1, 1, 1)$ and $\tilde{v}_1^- = (0, 0, 0)$ as shown in Table 7.9.

Stage 5: Calculate the distance of each alternative (d_i^{+}, d_i^{-}) from *FPIS* and *FNIS* using Eqs. 6.35 and 6.36, as shown below as an example illustrating this calculation.

Fig. 7.11 Membership functions of linguistic variable set

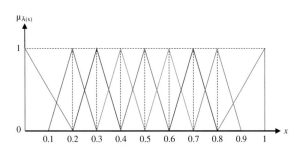

Table 7.8 Fuzzy decision matrix for nine alternatives with three criteria

Alternatives	C_1	C_2	C_3
A_1 (SR_1 DR_1 U_1 K_1)	(0.2, 0.3, 0.4)	(0.3, 0.4, 0.5)	(0.6, 0.7, 0.8)
A_2 (SR_1 DR_2 U_2 K_2)	(0.6, 0.7, 0.8)	(0.1, 0.2, 0.3)	(0.2, 0.3, 0.4)
A_3 (SR_1 DR_3 U_3 K_3)	(0.8, 1.0, 1.0)	(0.0, 0.0, 0.2)	(0.0, 0.0, 0.2)
A_4 (SR_2 DR_1 U_2 K_3)	(0.5, 0.6, 0.7)	(0.0, 0.0, 0.2)	(0.0, 0.0, 0.2)
A_5 (SR_2 DR_2 U_3 K_1)	(0.8, 1.0, 1.0)	(0.4, 0.5, 0.6)	(0.8, 1.0, 1.0)
A_6 (SR_2 DR_3 U_1 K_2)	(0.0, 0.0, 0.2)	(0.3, 0.4, 0.5)	(0.4, 0.5, 0.6)
A_7 (SR_3 DR_1 U_3 K_2)	(0.7, 0.8, 0.9)	(0.5, 0.6, 0.7)	(0.4, 0.5, 0.6)
A_8 (SR_3 DR_2 U_1 K_3)	(0.1, 0.2, 0.3)	(0.0, 0.0, 0.2)	(0.0, 0.0, 0.2)
A_9 (SR_3 DR_3 U_2 K_1)	(0.4, 0.5, 0.6)	(0.8, 1.0, 1.0)	(0.8, 1.0, 1.0)
Weight	0.497	0.295	0.208

Table 7.9 Weighted decisions for alternatives

Alternatives	C_1	C_2	C_3
A_1	(0.099, 0.149, 0.198)	(0.088, 0.118, 0.147)	(0.124, 0.145, 0.166)
A_2	(0.298, 0.347, 0.397)	(0.029, 0.059, 0.088)	(0.041, 0.062, 0.083)
A_3	(0.397, 0.497, 0.497)	(0.000, 0.000, 0.059)	(0.000, 0.000, 0.041)
A_4	(0.248, 0.298, 0.347)	(0.000, 0.000, 0.059)	(0.000, 0.000, 0.041)
A_5	(0.397, 0.497, 0.497)	(0.118, 0.147, 0.177)	(0.166, 0.208, 0.208)
A_6	(0.000, 0.000, 0.099)	(0.088, 0.118, 0.147)	(0.083, 0.104, 0.124)
A_7	(0.347, 0.397, 0.447)	(0.147, 0.177, 0.206)	(0.083, 0.104, 0.124)
A_8	(0.049, 0.099, 0.149)	(0.000, 0.000, 0.059)	(0.000, 0.000, 0.041)
A_9	(0.198, 0.248, 0.298)	(0.236, 0.295, 0.295)	(0.166, 0.208, 0.208)
A^+	$\tilde{v}_1^- = (1, 1, 1)$	$\tilde{v}_2^- = (1, 1, 1)$	$\tilde{v}_3^+ = (1, 1, 1)$
A^-	$\tilde{v}_1^- = (0, 0, 0)$	$\tilde{v}_2^- = (0, 0, 0)$	$\tilde{v}_3^+ = (0, 0, 0)$

Table 7.10 FTOPSIS results

Alternatives	d_j^+	d_j^-	CC_j	Rank
A_1	2.589	2.102	0.448	7
A_2	2.532	2.323	0.478	5
A_3	2.505	2.453	0.495	3
A_4	2.670	2.287	0.461	6
A_5	2.197	2.402	0.522	1
A_6	2.746	2.018	0.424	8
A_7	2.323	2.360	0.504	2
A_8	2.869	2.094	0.422	9
A_9	2.284	2.188	0.489	4

Stage 6: Calculate the closeness coefficient of each alternative using Eq. 6.39. The final results obtained by the FTOPSIS are summarized in Table 7.10 and presented in Fig. 7.12.

$$d_1^+ = \sqrt{\frac{1}{3}\left[(1-0.099)^2 + (1-0.149)^2 + (1-0.198)^2\right]}$$
$$+ \sqrt{\frac{1}{3}\left[(1-0.088)^2 + (1-0.118)^2 + (1-0.147)^2\right]}$$
$$+ \sqrt{\frac{1}{3}\left[(1-0.124)^2 + (1-0.145)^2 + (1-0.166)^2\right]} = 2.589$$
$$d_1^- = \sqrt{\frac{1}{3}\left[(0-0.099)^2 + (0-0.149)^2 + (0-0.198)^2\right]}$$
$$+ \sqrt{\frac{1}{3}\left[(0-0.088)^2 + (0-0.118)^2 + (0-0.147)^2\right]}$$
$$+ \sqrt{\frac{1}{3}\left[(0-0.124)^2 + (0-0.145)^2 + (0-0.166)^2\right]} = 2.102$$

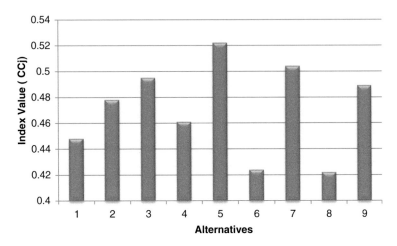

Fig. 7.12 Final ranking of alternatives using FTOPSIS

7.5 Analysis of Results and Discussion

After the results of the FDSS and the FAHP-FTOPSIS approaches are obtained, four steps are performed to analyze the results: first, comparing the obtained results from FDSS with those from FAHP-FTOPSIS to check the reliability of the proposed approach; second, verifying the obtained results using a sensitivity analysis; third, predicting the optimal combination of scheduling factors using analysis of mean; fourth, validating the obtained results via a confirmation test.

7.5.1 Comparison of the Results

In this section, a comparison between FDSS and FAHP-FTOPSIS results is presented by first analyzing the results obtained by FDSS and FAHP-FTOPSIS, and then comparing the outcomes of these two approaches.

Based on FDSS results in Table 7.4 and Fig. 7.9, it can be stated that the alternative A_5 (SR_2 DR_2 U_3 K_1) is the optimal combination of scheduling factors, with a *MPCI* value of 0.824. The alternative A_7 (SR_3 DR_1 U_3 K_2) is ranked second with a *MPCI* value of 0.723. The alternative A_9 (SR_3 DR_3 U_2 K_1) has a *MPCI* value of 0.676 and is ranked third, followed by the alternative A_3 (SR_1 DR_3 U_3 K_3) with a *MPCI* value of 0.50. The least desirable alternative is A_8 (SR_3 DR_2 U_1 K_3) with a *MPCI* value of 0.058. Thus, the descending order of the final ranking for the combination of scheduling factors using FDSS is: $A_5 > A_7 > A_9 > A_3 > A_2 > A_4 > A_1 > A_6 > A_8$.

Based on FAHP-FTOPSIS results, Table 7.10 and Fig. 7.12 show that the alternative A_5 (SR_2 DR_2 U_3 K_1) is closest to the fuzzy positive ideal solution (*FPIS*) with a value of 2.197, and the alternative A_8 (SR_3 DR_2 U_1 K_3) is located farthest from the fuzzy negative ideal solution (*FNIS*) with a value of 2.869. The alternative A_3 (SR_1 DR_3 U_3 K_3) is located farthest from the *FNIS* with a value of 2.453, and the alternative A_6 (SR_2 DR_3 U_1 K_2) is closest to the *FNIS* with a value of 2.018. According to the results in Table 7.10 and Fig. 7.12, it can be seen that the alternative A_5 (SR_2 DR_2 U_3 K_1) is the optimal combination of scheduling factors, with a relative closeness coefficient (*CC$_j$*) value of 0.522. The alternative A_7 (SR_3 DR_1 U_3 K_2) has a CC_j value of 0.504 and is ranked second in these results. The alternative A_3 (SR_1 DR_3 U_3 K_3) is ranked third with a CC_j value of 0.495. The worst alternative is A_8 (SR_3 DR_2 U_1 K_3) with a CC_j value of 0.422. Therefore, the descending order of the final ranking using FAHP-FTOPSIS is: $A_5 > A_7 > A_3 > A_9 > A_2 > A_4 > A_1 > A_6 > A_8$.

7.5.2   Sensitivity Analysis

To carry out the sensitivity analysis for the results obtained by FDSS, the fuzzy rules used to control the *MPCI* must be changed. This can be achieved by altering the fuzzy rules for different scenarios. Thus, Scenario 0 represents the *MPCI* values that are calculated by considering the fuzzy rules given in Table 7.4, while the other scenarios (Scenario 1, Scenario 2 and Scenario 3) show the *MPCI* values that are calculated by considering the altered fuzzy rules (Appendix C). A sensitivity analysis obtained by changing the fuzzy rules of the scheduling criteria is calculated as shown in Table 7.11, and graphically represented in Fig. 7.13.

From Table 7.11 and Fig. 7.13 it can be seen that the final ranking in Scenario 0 is relatively less sensitive to the changes in fuzzy rules. For example, A_5 has the highest value of *MPCI* in *Scenario 2* and *Scenario 3* and is the same as *Scenario 0*, while A_7 has the highest value of *MPCI* just in *Scenario 1*. The second rank goes to A_5 in *Scenario 1*, and A_7 in *Scenarios 2* and *3* has the same rank as in *Scenario 0*.

Table 7.11 Sensitivity analysis for the results obtained by FDSS

Alternatives	MPCI			
	Scenario 0	Scenario 1	Scenario 2	Scenario 3
A_1 (SR_1 DR_1 U_1 K_1)	0.330	0.313	0.325	0.304
A_2 (SR_1 DR_2 U_2 K_2)	0.419	0.504	0.528	0.504
A_3 (SR_1 DR_3 U_3 K_3)	0.500	0.667	0.523	0.667
A_4 (SR_2 DR_1 U_2 K_3)	0.383	0.442	0.492	0.550
A_5 (SR_2 DR_2 U_3 K_1)	0.824	0.824	0.833	0.833
A_6 (SR_2 DR_3 U_1 K_2)	0.267	0.157	0.212	0.246
A_7 (SR_3 DR_1 U_3 K_2)	0.723	0.833	0.793	0.793
A_8 (SR_3 DR_2 U_1 K_3)	0.058	0.058	0.058	0.058
A_9 (SR_3 DR_3 U_2 K_1)	0.676	0.676	0.610	0.676

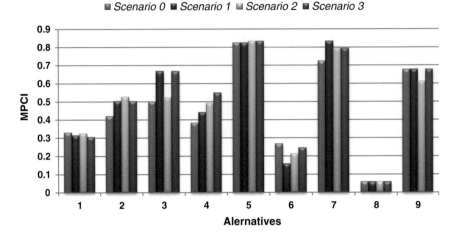

Fig. 7.13 Results of sensitivity analysis for FMCDM

A_8 has the lowest value of MPCI in all four scenarios, followed by A_6 and A_1. Thus, the sensitivity analysis reflects the stability of the decision outcome in *Scenario* 0, and verifies the robustness and the appropriateness of using the FDSS.

To conduct the sensitivity analysis for the results obtained by FAHP- FTOPSIS, two procedures are utilized. The first procedure is to increase or decrease the weight of each individual criterion, and keep the summation total of all criteria weights equal to one. The decision-maker can assume different scenarios. *Scenario* 0 shows the original weights' values as calculated by FAHP, whereas the other scenarios (Scenario 1, Scenario 2 … Scenario *n*) all show different weights values. For example, the total weight of the makespan (C_1), total tardiness (C_2) and number of tardy jobs (C_3) is equal to one ($0.497 + 0.295 + 0.208 = 1$); these original weights represent *Scenario* 0. The other scenarios can be established for example by increasing C_1 to 0.55, increasing C_2 to 0.30 and decreasing C_3 to 0.15. Table 7.12 summarizes the numerical results of three different scenarios. The second procedure is to exchange each criterion's weight with another criterion's weight. For instance, three possible Scenarios can be considered for C_1, C_2 and C_3, as shown in Table 7.13. The first Scenario is to exchange the weight of C_2 with C_3, while C_1 is constant. The second is to replace the weight of C_1 with C_2, while C_3 is constant. The third is to exchange the weight of C_1 with C_3 while C_2 is constant. According to the information of both procedures given in Tables 7.12 and 7.13, a sensitivity analysis, after changing the weights of scheduling criteria, is implemented as shown in Tables 7.14 and 7.15 (see Appendix D for detailed results). The results of the sensitivity test are also graphically plotted in Figs. 7.14 and 7.15.

As shown in Table 7.14 and Fig. 7.14, the final ranking in *Scenario* 0 is nearly the same as for all the other scenarios (*Scenario* 1, *Scenario* 2 and *Scenario* 3). For example, A_5, A_7, A_2, A_4 and A_1 have the same rankings in all scenarios. In this

Table 7.12 Increase/decrease the weight of individual criterion

Criteria	Scenario 0	Scenario 1	Scenario 2	Scenario 3
Makespan (C_1)	0.497	0.55	0.50	0.45
Total tardiness (C_2)	0.295	0.30	0.20	0.35
Number of tardy jobs (C_3)	0.208	0.15	0.30	0.20

Table 7.13 Exchange criterion's weight

Criteria	Scenario 0	Scenario 1	Scenario 2	Scenario 3
Makespan (C_1)	0.497	0.497	0.295	0.208
Total tardiness (C_2)	0.295	0.208	0.497	0.295
Number of tardy jobs (C_3)	0.208	0.295	0.208	0.497

Table 7.14 Sensitivity analysis for FAHP-FTOPSIS based on first procedure

Alternatives	Relative closeness coefficient (CCj)			
	Scenario 0	Scenario 1	Scenario 2	Scenario 3
A_1 ($SR_1\ DR_1\ U_1\ K_1$)	0.448	0.450	0.448	0.447
A_2 ($SR_1\ DR_2\ U_2\ K_2$)	0.478	0.485	0.479	0.473
A_3 ($SR_1\ DR_3\ U_3\ K_3$)	0.495	0.505	0.495	0.486
A_4 ($SR_2\ DR_1\ U_2\ K_3$)	0.461	0.468	0.461	0.456
A_5 ($SR_2\ DR_2\ U_3\ K_1$)	0.522	0.529	0.525	0.515
A_6 ($SR_2\ DR_3\ U_1\ K_2$)	0.424	0.423	0.422	0.424
A_7 ($SR_3\ DR_1\ U_3\ K_2$)	0.504	0.511	0.501	0.499
A_8 ($SR_3\ DR_2\ U_1\ K_3$)	0.422	0.424	0.421	0.420
A_9 ($SR_3\ DR_3\ U_2\ K_1$)	0.489	0.492	0.486	0.489

Table 7.15 Sensitivity analysis for FAHP-FTOPSIS based on second procedure

Alternatives	Relative closeness coefficient (CCj)			
	Scenario 0	Scenario 1	Scenario 2	Scenario 3
A_1 ($SR_1\ DR_1\ U_1\ K_1$)	0.448	0.448	0.443	0.441
A_2 ($SR_1\ DR_2\ U_2\ K_2$)	0.478	0.478	0.453	0.441
A_3 ($SR_1\ DR_3\ U_3\ K_3$)	0.495	0.494	0.458	0.440
A_4 ($SR_2\ DR_1\ U_2\ K_3$)	0.461	0.461	0.438	0.426
A_5 ($SR_2\ DR_2\ U_3\ K_1$)	0.522	0.524	0.492	0.484
A_6 ($SR_2\ DR_3\ U_1\ K_2$)	0.424	0.422	0.427	0.425
A_7 ($SR_3\ DR_1\ U_3\ K_2$)	0.504	0.501	0.482	0.463
A_8 ($SR_3\ DR_2\ U_1\ K_3$)	0.422	0.421	0.415	0.409
A_9 ($SR_3\ DR_3\ U_2\ K_1$)	0.489	0.487	0.487	0.477

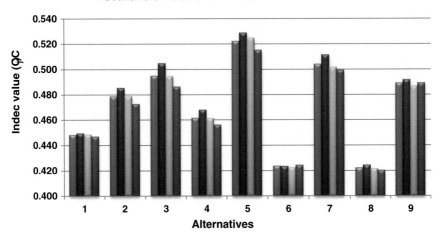

Fig. 7.14 Sensitivity analysis for FAHP-FTOPSIS based on first procedure

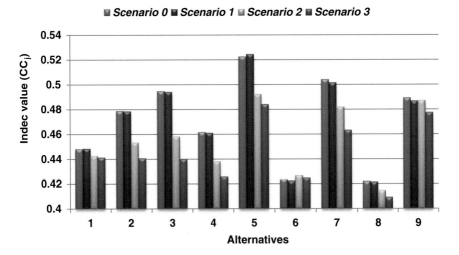

Fig. 7.15 Sensitivity analysis for FAHP-FTOPSIS based on second procedure

example, even though the weights of the criteria (C_1, C_2 and C_3) are each changed from the original weights, this has only a minor influence on the final ranking.

According to Table 7.15 and Fig. 7.15, it can be seen that the final ranking in *Scenario* 0 is also less sensitive in relation to the exchanges in the criteria weights with each other. For instance, A_5 is the first rank in all four Scenarios. A_7 is the second rank in *Scenario* 0, when the weight of C_2 and C_3 is exchanged. A_9 has the second rank in *Scenario* 2, when the weight of C_1 is replaced with C_2, and weight of C_1 and C_3 is exchanged. A_8 has the worst ranking, followed by A_6, in all Scenarios.

It can be concluded that almost all of the changes in the weight of each criterion do not have significant influence on the final ranking. Thus, the sensitivity analysis reflects the stability of the results in *Scenario* 0, and demonstrates the robustness of the FAHP-FTOPSIS approach.

7.5.3   Effect of Scheduling Factors on MPCI

The optimal scheduling factor can be assessed from the mean of each factor level. The effect of a factor level is the deviation it causes from the overall mean response. The mean of MCPI can be calculated using Eq. 6.3. The means of MPCI values are summarized in Table 7.16.

As shown in Table 7.16, it is clear that the optimal levels of SR (Sequencing Rule), DR (Dispatching Rule), U (Cell Utilization) and K (Due Date Tightness) are 2, 3, 3 and 1 respectively, due to their MPCI value.

The most significant scheduling factor can also be determined by calculating the difference between the maximum and minimum value for each factor. The greatest difference indicates the most significant factor. Moreover, it can be concluded that factor C (Cell Utilization) has the strongest effect on the scheduling of RFAC, followed by factor D (Due Date Tightness). It also can be seen that factor B (Dispatching Rule) has the weakest impact on the scheduling.

7.5.4   Confirmation Test

The final step is to predict and verify the improvement in the performance characteristics using the optimal levels of the scheduling factors. Based on Eq. 6.4,

Table 7.16 Response table of MPCI values

Factor	Applicable formula	MPCI	Max-Min	Rank
$A_{(1)}$	$(\eta_1 + \eta_2 + \eta_3)/3$	0.416	0.075	3
$A_{(2)}$	$(\eta_4 + \eta_5 + \eta_6)/3$	**0.491**		
$A_{(3)}$	$(\eta_7 + \eta_8 + \eta_9)/3$	0.486		
$B_{(1)}$	$(\eta_1 + \eta_4 + \eta_7)/3$	0.479	0.047	4
$B_{(2)}$	$(\eta_2 + \eta_5 + \eta_8)/3$	0.434		
$B_{(3)}$	$(\eta_3 + \eta_6 + \eta_9)/3$	**0.481**		
$C_{(1)}$	$(\eta_1 + \eta_6 + \eta_8)/3$	0.218	0.464	1
$C_{(2)}$	$(\eta_2 + \eta_4 + \eta_9)/3$	0.493		
$C_{(3)}$	$(\eta_3 + \eta_5 + \eta_7)/3$	**0.682**		
$D_{(1)}$	$(\eta_1 + \eta_5 + \eta_9)/3$	**0.610**	0.296	2
$D_{(2)}$	$(\eta_2 + \eta_6 + \eta_7)/3$	0.470		
$D_{(3)}$	$(\eta_3 + \eta_4 + \eta_8)/3$	0.314		

Table 7.17 Predicted results for initial and optimal conditions

	Initial condition	Optimal condition
Level	$A_2B_2C_3D_1$	$A_2B_3C_3D_1$
MPCI prediction	0.825	0.872
MPCI confirmed	0.824	0.871
PPE (%)	0.001	0.001
Improvement MPCI = 5.7 %		

the predicted *MPCI* for initial and optimal levels of scheduling factors can then be calculated as shown below:

For the initial levels, as mentioned earlier, $SR_2DR_2U_3K_1$ are selected to be the initial condition for the scheduling factors: $MPCI_{Initial} = 0.464 + (0.491 - 0.464) + (0.434 - 0.464) + (0.682 - 0.464) + (0.610 - 0.464) = 0.825$.

$MPCI_{confirmed}$ is also determined for both the initial and optimal levels of the scheduling factors. Table 7.4 shows that the alternative A_5 provides the largest MPCI value which represents $MPCI_{Initial}$, while $MPCI_{optimal}$ can be graphically calculated by obtaining the value of the optimal levels using the fuzzy logic rule.

The results of the confirmation run for the initial and optimal levels of the scheduling factors are summarised in Table 7.17. The results indicate that the optimal levels of scheduling factors produce the best *MPCI* in all trials. In other words, it can be seen that the MPCI value of the optimal level of scheduling factors is improved by approximately 0.11 compared to the initial level of scheduling factors. The results also show that the relative percentage deviation (PPE) is acceptable (less than 10 %). In summary, it can be said that the experimental results are confirmed.

7.6 Concluding Remarks

This chapter is devoted to the application of the developed method in Chap. 6, in order to demonstrate its capability in tackling real-world MCDM problems. The results achieved in this chapter certify that:

- The integrated approach is efficient and effective for finding the optimal dynamic scheduling of RFAC with fewer experiments compared to full factorial experimental methods.
- The optimum values of the four different factors have been found to be: TLPT (Sequencing rule), SNQ (Dispatching rule), 95 % (Cell Utilization) and 6 (Due date tightness).

- The most significant factors affecting the scheduling strategy have been identified as factor U (Cell Utilization) and factor K (Due Date Tightness), which account for nearly 92 %.
- A percentage predicate error of less than 10 % has been obtained. This percentage indicates that the suggested hybrid approach can be successfully used in the multi-objective optimization of scheduling in RFAC.
- The confirmed *MPCI* value of optimal scheduling factors produces the best result (0.871) against the initial setting (0.84). Therefore, the *MPCI* increased by nearly 6 % compared to the initial setting.

In summary, the obtained results (this chapter) showed that the integration of a fuzzy MCDM approach (Chap. 6) with Taguchi experimental design and simulation modelling (Chap. 5) has many valuable features: (1) it has the ability to examine the behavior of RFAC under different scheduling factors, (2) it addresses the problem of dynamic scheduling using fewer experiments compared to full factorial experimental methods, (3) it finds the optimal or near-optimal combination of the selected factors to optimize the multi-objectives simultaneously, and (4) it predicts the significant factors that affect the system performance.

Chapter 8
Conclusions and Recommendations for Future Work

8.1 Introduction

Flexible manufacturing systems have attracted significant attention in recent years due to their flexibility and capacity to deal with unexpected events. One class of such systems is called robotic flexible assembly cells (RFAC). The design of RFAC with more than one robot offers many advantages, e.g. efficiency due to a reduced work environment (Mohamed et al. 2001), increased robustness enabling the assembly of a variety of products using the same resources (Marian et al. 2003), and flexibility due to superior ease of modification and reconfiguring (Makino 1989). Accordingly, employing multi-robots in RFAC offers the advantage of increased productivity in a shorter cycle time with lower production costs. Nevertheless, there are certain difficulties that have arisen with this design concept. For example, two robots operating simultaneously in the same work environment require a complex control system to prevent collisions between them. Also, industrial robots must be employed as effectively as possible due to the high cost of the robots. To overcome these difficulties, efficient scheduling of RFAC is required.

Few studies have been done on the problem of scheduling RFAC. These studies may be categorized into three groups. The first group applied heuristic methods, while the second group investigated simulation approaches and the third group implemented expert systems to solve scheduling problems in RFAC. In these studies, three limitations were identified: (1) scheduling of RFAC only in a *single-product* assembly environment was considered; (2) scheduling of RFAC in a *static situation* was investigated; (3) scheduling of RFAC only in *single-objective* optimization problems was examined. In light of these limitations, this thesis presents new strategies for scheduling RFAC in a *multi-product* assembly environment, in which *dynamic status* and *multi-objective* optimization problems occur.

© Springer International Publishing Switzerland 2016
K.K. Abd, *Intelligent Scheduling of Robotic Flexible Assembly Cells*,
Springer Theses, DOI 10.1007/978-3-319-26296-3_8

The purpose of this final chapter is to:

- Summarize the research presented in this thesis.
- Describe the contributions of this research.
- Highlight the potential directions for future research.

8.2 Summary of the Research

The main theme of this research was solving scheduling problems in RFAC. In the course of the research, three challenges were faced. The first challenge was the scheduling of RFAC in a multi-product assembly environment. The second challenge was the scheduling of RFAC in a dynamic situation. The third challenge was the optimization of dynamic scheduling for multi-objective problems. Solutions to these three challenges had not been previously considered by researchers.

The three challenges were overcome using a combination of advanced solution approaches such as simulation modelling, artificial intelligence, Taguchi optimization method and statistical analysis tools. The details of the proposed solutions to these challenges, the results and the findings are briefly summarized in the following sub-sections.

8.2.1 Scheduling RFAC in a Multi-product Assembly Environment

In Chap. 3, a new methodology for scheduling RFAC in a multi-product assembly environment was developed. The developed methodology was divided into three modules: pre-processing, scheduling and simulation. In Chap. 4, the methodology developed in Chap. 3 was applied to a scenario-based case study of RFAC. The simulation results showed that the performance of the developed methodology was more efficient compared to existing scheduling rules such as SPT, LPT, RAND, EDD, CR and MST. This is because the developed methodology was built by combining all input variables such as processing time, due date and batch size. The simulation results also revealed that the existing scheduling rules were not suitable for finding an acceptable schedule for multi-objective criteria. The reason for the poor performance of existing scheduling rules was that these rules were not able to incorporate all the variables for the complete set of jobs.

However, although the methodology developed (Chap. 3) and implemented (Chap. 4) was devoted to scheduling RFAC in a multi-product assembly environment, it concentrated only on the *static scheduling* problem. Consequently, the following chapter (Chap. 5) expanded the developed methodology by considering the *dynamic scheduling* problems.

8.2.2 Scheduling RFAC in a Dynamic Situation

In Chap. 5, an intelligent approach for scheduling RFAC in a dynamic situation was developed. In this approach, the simulation modelling was integrated with Taguchi optimization method to study the influence of the scheduling factors on the performance of RFAC. Based on the simulation results and the analysis, the most significant scheduling factor in term of the effects on the RFAC performance was cell utilization (U) when considering time-based objectives such as the makespan (C_{max}). The due date tightness factor (K) was the most significant when due date-based objectives such as the total tardiness (TD) or number of tardy jobs (N_T) were considered. When the three objective functions were considered individually, the recommended best parameters for scheduling RFAC were as follows: the factor/level combination $SR_1DR_1U_3K_3$ for the C_{max}, $SR_3DR_3U_2K_1$ for the TD, and $SR_3DR_3U_3K_1$ for the N_T.

Although the developed approach was designed to deal with dynamic scheduling in RFAC, it was restricted to solving only single-objective optimisation problems. However, the generalised problem, without this restriction, was solved in Chap. 6.

8.2.3 Scheduling RFAC in Multi-objective Optimization Problems

In Chap. 6, an optimization approach to deal with multi-objective problems for the dynamic scheduling of RFAC was developed using fuzzy Multi-Criteria Decision-Making (MCDM) methods. There are two reasons for using MCDM methods within a fuzzy environment: first, the fuzzy MCDM method can effectively handle any complex decision problems; second, the MCDM method together with fuzzy set theory can deal with the decision problems when the information is uncertain and ambiguous.

In Chap. 7, the hybrid fuzzy MCDM approach described in Chap. 6 was examined, verified and validated, using a realistic case study where three phases were used to implement the approach developed: problem description; application of fuzzy MCDM; and analysis of the results. Different analysis tools, such as sensitivity analysis (SA) and percentage prediction error (PPR) were used. A confirmation test was also performed to check whether any improvement in the results might be obtained. The analysis showed the stability of the results for both FDSS and FAHP-FTOPSIS and also verified the robustness and the appropriateness of the proposed approach. Therefore, it was demonstrated that the hybrid fuzzy MCDM approach can be successfully used in the multi-objective optimization of dynamic scheduling in RFAC.

8.3   Conclusions

This research has contributed significantly towards solving scheduling problems of RFAC, using simulation modelling, fuzzy logic, Taguchi method and statistical analysis tools. The key contributions of this research are summarized as follows:

C_1 Development of a new methodology for scheduling RFAC in a multi-product assembly environment. This methodology incorporated all features from previous studies of scheduling RFAC in that these studies concentrated on assembling only one type of product at a time only. The noteworthy feature of the developed methodology is the combining of a fuzzy-based mathematical model with simulation modelling. This combination can be used to generate the schedule for assembling multi-products and to construct the simulation model of the RFAC.

C_2 Formulation of a model of the scheduling problem in RFAC using five objective functions, namely makespan, percentage of robots idle time, total tardiness, maximum tardiness and percentage of tardy jobs. The first two objectives were categorized as time based objectives, and the other three were categorized as due date based objectives. Hence, the formulated mathematical model is comprehensive for evaluating the RFAC performance under different scheduling policies.

C_3 Development of a new and sophisticated scheduling rule, namely fuzzy sequencing rule (FSR). The developed rule was constructed using a fuzzy-based mathematical model. This rule overcomes the deficiencies in the existing scheduling rules by considering all the important input variables in the scheduling problems such as processing time, batch size, due date and number of required stations. In FSR, the membership functions were used to find the contribution of each product type to the output, and then to generate the sequence of products flow to the RFAC. The FSR is also validated through a realistic case study of RFAC. The simulation results demonstrated that the FSR was more efficient compared to the heuristic scheduling rules, when considering the objectives of minimizing makespan, robots idle time, total tardiness, maximum tardiness and number of tardy jobs. Moreover, the simulation results confirmed what some previous studies had already highlighted, that the heuristic scheduling rules were not able to find an acceptable schedule regarding multi-objective criteria.

C_4 Development of a novel methodology for scheduling RFAC in a dynamic environment. In the developed methodology, Taguchi experimental design method and simulation modelling are used as tools. This methodology has many valuable features: (1) it has the ability to examine the behavior of RFAC under different scheduling factors, (2) it addresses the problem of scheduling using significantly fewer experiments compared to full factorial experimental methods, (3) it finds the optimal or near-optimal combination of the selected

factors that optimize the objective functions and (4) it predicts the significant
factors that affect the system performance.

C_5 Integration of FAHP (Fuzzy Analytic Hierarchy Process) with FTOPSIS
(Fuzzy Technique for Order Preference by Similarity to Ideal Solution) for
optimizing the dynamic scheduling in RFAC under different performance
measures. This integration offers significant advantages such as dealing with
complex decision problems, and handling any decision problems when the
information is uncertain due to vagueness and imprecision. In this integration,
the FAHP was applied to determine the relative weights for multi scheduling
criteria and the FTOPSIS was applied for evaluation of each alternative
scheduling combination based on its overall performance in order to make a
final decision.

C_6 Development of a Fuzzy Decision Support System (FDSS) which is able to
find the optimal solution for multi-objective problems in dynamic scheduling
of RFAC. The FDSS was built, using Matlab fuzzy logic toolbox, to derive
the optimal solution. The FDSS has the ability to combine multi-objective
functions in one performance measure named multiple performance characteristics index (MPCI). The main advantage of this system is the capacity to
mimic human expert reasoning for optimizing the dynamic scheduling in
RFAC, and to also deal with imprecise and uncertain information.

C_7 Validation of the application of the fuzzy decision support system (FDSS) and
fuzzy AHP-fuzzy TOPSIS (FAHP-FTOPSIS). This validation was accomplished by checking the reliability and consistency of the obtained results of
the proposed FDSS and FAHP-FTOPSIS, and then performing a sensitivity
analysis to verify the final results obtained with FDSS and FAHP-FTOPSIS.

8.4 Recommendations for Future Work

The presented study for the scheduling of RFAC is expected to open new areas for
future research. There are several directions which can be recommended to continue
this work. The potential directions for further research can be classified into three
categories: (1) the development of robust scheduling of RFAC with interruptions
(2) the application of virtual reality for RFAC simulation (3) the development of
deadlock prevention and avoidance policies for the RFAC. The following
sub-sections summarize the significance of research in these directions.

8.4.1 Robust Scheduling of RFAC with Interruptions

The current research took into consideration those unexpected events that can be
categorized as job-related, such as due date changing and early or late arrival time

of jobs. These events may cause deviations from the generated schedules. An interesting area of future work is to expand the current research study to be more applicable to real scheduling problems, by taking into account other influential events that can be categorized as resource-related, such as robot breakdowns during the scheduling of RFAC.

Three uncertain parameters may be considered within the scope of this area of study when the interruptions on the shop floor (robot breakdowns) occur: (1) the number of robots subject to breakdown and repair; (2) breakdown frequency, which represents how many times each robot will break down during the scheduling horizon; (3) repair duration, which represents the mean time required to repair the robot after its breakdown. The values of the above uncertain parameters can be described by several known distributions such as a normal distribution, an exponential distribution or a uniform distribution.

8.4.2   Virtual Reality for RFAC Simulation

Virtual reality is one of the most attractive research areas for providing a deeper understanding of the behavior of RFAC. Virtual reality tools would significantly improve the scheduling presentation in a dynamic environment. To date, a number of virtual tools, such as RobotStudio (Qi et al. 2008; Zhang and Qi 2008), RoboCAD (Aguiar et al. 2008), IGRIP (Cheng 2003) and Workspace (Ahrens and Pageau 2002), have become available, especially for visualizing a robotic cell. Some of the virtual tools have been developed by universities and educational institutions and the others by industrial companies. These tools supply a highly-detailed 3D layout of a robotic cell and also help to check and evaluate several parameters, such as collisions detection, optimal path, and estimate of cycle time. In future research, Robostudio would be recommended to be used as it has several features that make it a crucial tool for robot programmers. Robotstudio allows offline programming of the system without interrupting production, decreases risk by visualizing and validating solutions and layouts, allows programming of new activities without interrupting production. It also has the ability to import data from CAD in several formats such as IGES, STEP and VDAFS, and allows automatic detection of collision of the robot with the system equipment. Simulation of RFAC in a real world context will be accomplished through two stages, modelling and layout design.

- Modelling of RFAC, where solid models of assembly stations, material handling devices and gripper changing station are constructed. During this stage, peripheral equipment such as tools and fixtures are placed into the simulated RFAC.
- RFAC layout, where the working envelope of the robots will be determined, and the offline programming (OLP) of the RFAC will be generated. The goal of this stage is to ensure that the robots' arms will move without collision.

8.4.3   Deadlock Prevention and Avoidance in RFAC

In the current research, more than one type of product is processed concurrently in RFAC. The assembly sequences of these products were executed simultaneously via a set of shared resources such as robots, assembly stations, parts feeder, gripper changing station, fixtures, and input and output conveyors. In such a system, the sharing of resources may lead to a deadlock which is a highly undesirable situation. Hence, one of the challenging problems in RFAC operation is to assign shared resources to different types of products efficiently and without causing deadlocks.

Deadlock problems can be solved simultaneously at both scheduling and controlling stages. One way of dealing with deadlock problems is to model any flexible manufacturing systems with petri nets technique (Li and Zhou 2009). An interesting area of research will be to model RFAC using petri nets to guarantee obtaining deadlock-free schedules.

8.5   Final Word

The research presented in this thesis has demonstrated the capability of overcoming the limitations of previous studies of scheduling robotic flexible assembly cell (RFAC). Three major improvements were achieved in this research. The first improvement was the development of a new methodology for scheduling RFAC in a *multi-product assembly* environment. The proposed methodology was applied to a scenario-based case study of RFAC. The simulation results showed that the suggested methodology was more efficient compared to the existing scheduling policies. The second improvement was developing an intelligent approach for scheduling RFAC in a *dynamic situation*. This approach combined advanced solution approaches to evaluate the RFAC performance under different scheduling policies. Then, statistical analysis tools were employed to determine the most significant scheduling factors which affect the system performance. The third improvement was extending the developed approach to optimize the dynamic scheduling in RFAC for *multi-objective problems*. The results obtained showed that the proposed methodology has the ability to tackle any complex decision problems where the information is uncertain and ambiguous. The analysis tools confirmed that the proposed approach can be effectively applied to optimize dynamic scheduling in RFAC.

The above three improvements were achieved using a combination of simulation modelling, artificial intelligence, Taguchi optimization method, fuzzy MCDM techniques and statistical analysis tools. As per the literature review, these combinations have not previously been applied to the scheduling problems in RFAC. In addition, there is no documented research that indicates the application of both FDSS and FAHP-FTOPSIS for optimization of dynamic scheduling problems.

Even though three key improvements for the scheduling of RFAC were demonstrated in this thesis, possible improvements and future directions were recommended in the previous section to continue this research and to make it even more applicable to real life situations.

References

Aguiar, A. J. C., Villani, E., & Junqueira, F. (2008). Graphic robot simulation for the design of work cells in the aeronautic industry. *ABCM Symposium Series in Mechatronics, 3*, 346–354.

Ahrens, G., & Pageau, G. (2002). Trends in the robotic simulation industry. *Assembly Automation, 22*(3), 230–234.

Cheng, F. S. A. (2003). The simulation approach for designing robotic workcells. *Journal of Engineering Technology, 20*(2), 42–48.

Li, Z. W., & Zhou, M. C. (2009). *Deadlock resolution in automated manufacturing systems: A novel petri net approach.* London: Springer.

Makino, H., (1989). The comparison between robotic assembly line and robotic assembly cell. In *Proceedings of the 10th International Conference on Assembly Automation* (pp. 1–10). Canazawa, Japan.

Marian, R. M., Kargas, A., Luong, L. H. S., & Abhary, K. (2003). 'A framework to planning robotic flexible assembly cells'. In *32nd International Conference on Computers and Industrial Engineering* (pp. 607–615). Limerick, Ireland.

Mohamed, S. B., Petty, D. J., Harrison, D. K., & Rigby, R. (2001). A cell management system to support robotic assembly. *The International Journal of Advanced Manufacturing Technology, 18*(8), 598–604.

Qi, L., Yin, X., Wang, H., & Tao, L. (2008). Virtual engineering: Challenges and solutions for intuitive offline programming for industrial robot. In *Proceedings of the IEEE International Conference on Robotics, Automation and Mechatronics* (pp. 12–17). Chengdu, China.

Zhang, D., & Qi, L. (2008). Virtual engineering: optimal cell layout method for improving productivity for industrial robot. In *Proceedings of the IEEE International Conference on Robotics, Automation and Mechatronics* (pp. 6–11).

Appendix A

Examples of configurations of robotic flexible assembly cells (Figs. A.1, A.2, A.3 and A.4).

Fig. A.1 Robots assembly cell (Pelagagge et al. 1995)

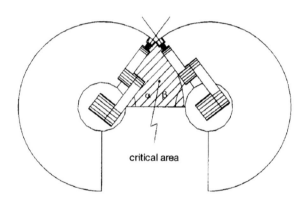

Fig. A.2 A two-robot assembly cell (Jiang et al. 1998)

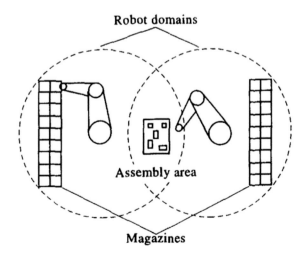

© Springer International Publishing Switzerland 2016
K.K. Abd, *Intelligent Scheduling of Robotic Flexible Assembly Cells*,
Springer Theses, DOI 10.1007/978-3-319-26296-3

Fig. A.3 A two-robot assembly cell (Lee and Lee 2002)

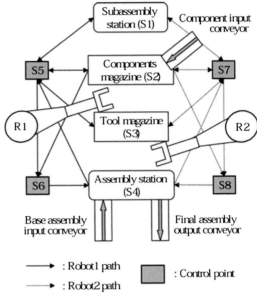

Fig. A.4 A robotic flexible assembly cell (Marian et al. 2003)

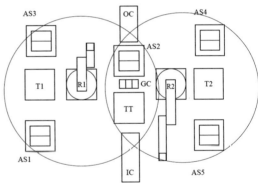

References

Jiang, K., Seneviratne, L. D., & Earles, S. W. E. (1998). Scheduling and compression for a multiple robot assembly workcell. *Production Planning & Control: The Management of Operations. 9*(2), 143–154.

Lee, J. K., & Lee, T.E. (2002). Automata-based supervisory control logic design for a multi-robot assembly cell. *International Journal of Computer Integrated Manufacturing. 15*(4), 319–334.

Marian, R. M., Kargas, A., Luong, L.H. S., & Abhary, K. (2003). A framework to planning robotic flexible assembly cells. *32nd International Conference on Computers and Industrial Engineering* (pp. 607–615). Limerick, Ireland.

Pelagagge, P. M., Cardarelli, G., & Palumbo, M. (1995). Design criteria for cooperating robots assembly cells. *Journal of Manufacturing Systems. 14*(4), 219–229.

Appendix B

SIMPROCESS activities that are used to build and simulate the assembling processes in RFAC.

Icon	Element	Description
	Generate	Produces entities according to a selected format, typically according to a selected statistical distribution, but uses no resources
	Dispose	Disposes of an entity, using no resources and requiring no simulation time
	Get resource	Provides a mechanism for capturing resources such as assembly station that may be used for a number of downstream activities
	Free resource	Provides a mechanism for releasing resources that were captured by a GET RESOURCE activity
	Delay	Represents any activity during which an entity is delayed for a specified time in its progress through the model (system)
	Assemble	Involves the combining of listed entities into a new output entity. For example, the development of a business proposal may contain three documents that are merged using an assembly activity
	Branch	Allows for defining alternative routings for flow objects. Branching may be based on a probability or a condition. For example, the outcome of an inspection process may be modeled using probabilistic branching
	Merge	Provides a mechanism for merging a number of connectors into a single connector
	Gate	Holds entities in a queue, until a signal is received. For example, a GATE activity would be used to model orders held in inventory until a signal is received from the distributor to fulfil the demand

(continued)

© Springer International Publishing Switzerland 2016
K.K. Abd, *Intelligent Scheduling of Robotic Flexible Assembly Cells*,
Springer Theses, DOI 10.1007/978-3-319-26296-3

Icon	Element	Description
	Join	Takes the clones and original entity that were split up, and matches them to produce the original one. For example, a JOIN activity may be used for matching the paperwork with the shipment
	Assign	Provides a mechanism for defining or changing attributes values
	Transfer	Routes entities from one portion of a model to another without using a connector, or routes entities to another model

Appendix C

Fuzzy rules and surface plots of the inputs/output for three scenarios (Table C.1; Fig. C.1).

Table C.1 Fuzzy rules under different scenarios

S/N ratio for			MPCI		
S/N of C_{max}	S/N of TD	S/N of N_{TD}	Scenario 1	Scenario 2	Scenario 3
Low	Low	Low	Tiny	Tiny	Tiny
		Medium	Tiny	Tiny	Tiny
		High	Very Small	Very Small	Tiny
	Medium	Low	Very Small	Very Small	Very Small
		Medium	Very Small	Small	Very Small
		High	Very Small	Small	Very Small
	High	Low	Small	Small	Very Small
		Medium	Small	Small	Very Small
		High	Small	Medium	Small
Medium	Low	Low	Small	Medium	Small
		Medium	Small	Medium	Small
		High	Small	Medium	Medium
	Medium	Low	Medium	Medium	Medium
		Medium	Medium	Medium	Medium
		High	Medium	Medium	Medium
	High	Low	Large	Large	Medium
		Medium	Large	Large	Large
		High	Large	Large	Large

(continued)

© Springer International Publishing Switzerland 2016
K.K. Abd, *Intelligent Scheduling of Robotic Flexible Assembly Cells*,
Springer Theses, DOI 10.1007/978-3-319-26296-3

Table C.1 (continued)

S/N ratio for			MPCI		
S/N of C_{max}	S/N of TD	S/N of N_{TD}	Scenario 1	Scenario 2	Scenario 3
High	Low	Low	Large	Large	Large
		Medium	Large	Large	Very Large
		High	Large	Very Large	Very Large
	Medium	Low	Very Large	Very Large	Very Large
		Medium	Very Large	Very Large	Very Large
		High	Very Large	Very Large	Very Large
	High	Low	Very Large	Huge	Huge
		Medium	Huge	Huge	Huge
		High	Huge	Huge	Huge

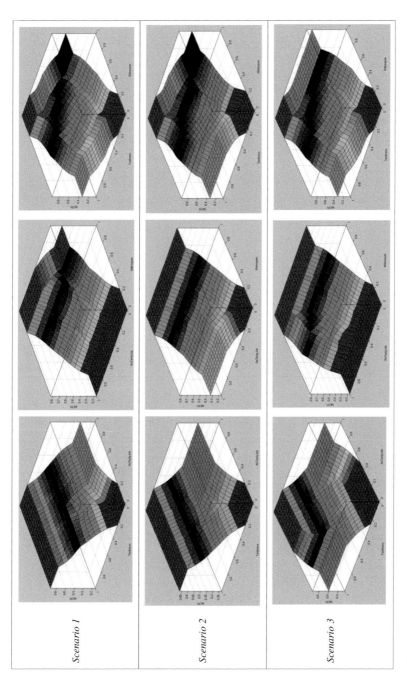

Scenario 1

Scenario 2

Scenario 3

Fig. C.1 3D surface plots of the inputs/output under different scenarios

Appendix D

Results of the sensitivity analysis for FAHP-FTOPSIS (Tables D.1, D.2, D.3, D.4, D.5 and D.6).

Table D.1 The results of increasing/decreasing the weight of individual criteria—scenario 1

Alternatives	d_j^+	d_j^-	cc_j	Rank
A_1	2.612	2.133	0.449522	5
A_2	2.512	2.368	0.485244	7
A_3	2.460	2.506	0.504643	3
A_4	2.642	2.323	0.467889	9
A_5	2.200	2.469	0.528776	2
A_6	2.770	2.034	0.423436	4
A_7	2.307	2.413	0.511224	1
A_8	2.862	2.109	0.424281	8
A_9	2.307	2.231	0.49165	6

Table D.2 The results of increasing/decreasing the weight of individual criteria—scenario 2

Alternatives	d_j^+	d_j^-	cc_j	Rank
A_1	2.562	2.082	0.448414	5
A_2	2.522	2.314	0.478524	7
A_3	2.503	2.449	0.494627	3
A_4	2.668	2.283	0.461096	9
A_5	2.156	2.380	0.524687	2
A_6	2.738	2.002	0.422281	4
A_7	2.332	2.346	0.501497	1
A_8	2.868	2.088	0.421341	6
A_9	2.285	2.164	0.486419	8

© Springer International Publishing Switzerland 2016
K.K. Abd, *Intelligent Scheduling of Robotic Flexible Assembly Cells*,
Springer Theses, DOI 10.1007/978-3-319-26296-3

Table D.3 The results of increasing/decreasing the weight of individual criteria—scenario 3

Alternatives	d_j^+	d_j^-	cc_j	Rank
A_1	2.586	2.090	0.446874	5
A_2	2.557	2.291	0.47263	7
A_3	2.546	2.409	0.486211	9
A_4	2.695	2.259	0.456032	3
A_5	2.221	2.360	0.51524	2
A_6	2.732	2.014	0.424352	4
A_7	2.332	2.324	0.499121	1
A_8	2.875	2.084	0.420269	6
A_9	2.264	2.166	0.488995	8

Table D.4 The results of exchanging criteria weights—scenario 1

Alternatives	d_j^+	d_j^-	cc_j	Rank
A_1	2.563	2.083	0.448319	5
A_2	2.524	2.313	0.478165	7
A_3	2.505	2.447	0.494116	3
A_4	2.670	2.282	0.460786	9
A_5	2.160	2.379	0.524131	2
A_6	2.738	2.002	0.422391	4
A_7	2.332	2.344	0.501324	1
A_8	2.869	2.088	0.42127	6
A_9	2.284	2.164	0.486535	8

Table D.5 The results of exchanging criteria weights—scenario 2

Alternatives	d_j^+	d_j^-	cc_j	Rank
A_1	2.569	2.040	0.442589	5
A_2	2.633	2.181	0.453036	9
A_3	2.680	2.263	0.45787	7
A_4	2.778	2.165	0.438017	3
A_5	2.284	2.213	0.492066	2
A_6	2.679	1.994	0.426729	1
A_7	2.364	2.197	0.481793	4
A_8	2.896	2.050	0.414513	6
A_9	2.197	2.085	0.486941	8

Table D.6 The results of exchanging criteria weights—scenario 3

Alternatives	d_j^+	d_j^-	cc_j	Rank
A_1	2.473	1.952	0.441105	5
A_2	2.648	2.085	0.440549	9
A_3	2.755	2.163	0.439842	7
A_4	2.824	2.094	0.425769	1
A_5	2.197	2.062	0.484104	2
A_6	2.621	1.936	0.424757	3
A_7	2.410	2.079	0.463133	4
A_8	2.907	2.013	0.409115	6
A_9	2.160	1.972	0.477302	8

Curriculum Vitae

Khalid Karam Abd
University of South Australia
e-mail: khalid.abd@mymail.unisa.edu.au
http://people.unisa.edu.au/khalid.abd
Mobile: +61402440314,
Office: +61883025368

Education	
2010–2014	University of South Australia, Adelaide, Australia: PhD in Advanced Manufacturing Engineering. Thesis title *Intelligent Scheduling of Robotic Flexible Assembly Cells*
1999–2002	University of Technology, Baghdad, Iraq: M.Sc. in Industrial Engineering. Thesis title *Computer Aided Planning of Assembly for Mechanical Products*
1995–1999	University of Technology, Baghdad, Iraq: B.Sc. in Production Engineering
Employment History	
2011–2015	Lecturer and Tutor at University of South Australia
2006–2009	Lecturer at University of Technology, Baghdad-Iraq
2003–2006	Assistant lecturer at University of Technology, Baghdad-Iraq
Work Experience	
2014–current	**Lecturer** at University of South Australia. Lectured the following courses: Engineering Dynamic and Engineering Maintenance
2011–current	**Tutor** at University of South Australia. Tutored the following courses: Intelligent Manufacturing Systems, Supply Chain Management, Engineering Dynamic and Intelligent Production Systems

(continued)

© Springer International Publishing Switzerland 2016
K.K. Abd, *Intelligent Scheduling of Robotic Flexible Assembly Cells*,
Springer Theses, DOI 10.1007/978-3-319-26296-3

2003–2009	**Assistant Lecturer/Lecturer** at University of Technology. Lectured the following courses: Engineering Drawing, Descriptive Geometry and Manufacturing Processes

Special Awards

2013	Young Scientist Award, presented by the Organizing Committee of the International Conference 2013, Adelaide, South Australia
2012	Best Poster Award, 2nd place in 2012
2011	Best Poster Award, 1st place in 2011

Printed in the United States
By Bookmasters